BrightRED Study Guide

Curriculum for Excellence

N4

MATHEMATICS

Brian J Logan

First published in 2014 by:
Bright Red Publishing Ltd
1 Torphichen Street
Edinburgh
EH3 8HX

A CIP record for this book is available from the British Library

ISBN 978-1-906736-50-7

With thanks to:
PDQ Digital Media Solutions Ltd (layout), Clodagh Burke (edit)

Cover design by Caleb Rutherford – e i d e t i c

Acknowledgements
Every effort has been made to seek all copyright holders. If any have been overlooked, then
Bright Red Publishing will be delighted to make the necessary arrangements.

Permission has been sought from all relevant copyright holders and Bright Red Publishing are
grateful for the use of the following:
Beefypeeg/freeimages.com (page 8); manahan/freeimages.com (page 8); encrier/iStock.com
(page 10); Ridofranz/iStock.com (page 17); smudgerone (page 19); tilo/iStock.com (page 22);
ollycb/freeimages.com (page 22); Alan Levine (CC BY-SA 2.0 https://creativecommons.org/
licenses/by-sa/2.0/) (page 23); brokenarts/freeimages.com (page 24); ba1969/freeimages.
com (page 24); nazreth/freeimages.com (page 24); just1999/freeimages.com (page 28);
MartyIOM (public domain) (page 28); Scott Waldron/iStock.com (page 42); Gringer (public
domain) (page 43); Cdang (public domain) (page 48); cobrasoft/freeimages.com (page 58);
Picsfive/iStock.com (page 59); GordonBellPhotography/iStock.com (page 68); Aaron Amat/
Shutterstock.com (page 72).

Printed and bound in the UK by Charlesworth Press

CONTENTS

NATIONAL 4 MATHEMATICS STUDY GUIDE

NATIONAL 4 MATHEMATICS

INTRODUCTION TO NATIONAL 4 MATHEMATICS

PREFACE

Welcome to this study guide for National 4 Mathematics. The fact that you are reading the guide shows that you care about your performance in the course and proves that you are determined to do well and hopefully gain the required qualification.

In order to be successful at National 4 Mathematics, you will have to prepare properly. You must attend lessons, study the course, practise key examples and ask for advice about areas of concern. Special preparation will be necessary during the time before assessments. This guide is designed to help you achieve your aims, but it is not enough on its own. You can ask your teacher or lecturer for advice, discuss the course with friends and possibly attend study classes.

HOW TO USE THIS BOOK

The first three chapters in the book cover the National 4 course. These chapters cover the three units of the course, Expressions and Formulae, Relationships and Numeracy. There are 40 two-page spreads covering the major aspects of National 4 Mathematics. It is possible that you will be taught topics in a different order from here, although this is the order in which the units appear on the Scottish Qualifications Authority (SQA) website. You may, for example, decide that it is best to start with the chapter on Numeracy as this contains many important themes which run through the first two chapters such as whole numbers, rounding, decimals and fractions. It is up to you to decide the order of study. You can organise your study to fit in with your class lessons and then use the guide to help prepare for your assessments.

The guide also contains a complete practice added value unit test with detailed worked solutions. There is a spread containing more practice covering a selection of questions from each unit. Check these sections before your assessments.

Each spread contains the key elements from each topic. There are 'Don't Forget' hints which include vital areas to focus on within each spread.

Each spread contains advice, formulae, diagrams and examples of exam standard with solutions and hints for every topic.

Each spread will end with a section called 'Things to do and think about' which will contain examples for you to practise. There will be solutions to all the examples in these sections and solutions to the 'More practice' section too. Another innovation in the book will be a course idea for most of the 40 spreads. These will suggest ideas for extra study and are intended to improve your mathematical skills and knowledge.

As you work through the book, try the examples *before* you read the solutions if possible. Remember that this book on its own is not enough to ensure that you are successful at National 4 Mathematics. Success also requires regular attendance at class, hard work and proper preparation. If there is an area of the syllabus that you feel unsure of, address the situation and ask for appropriate help.

Do not leave things to the last minute when studying, and do not try to do too much at the one time. The way to get the maximum benefit from the guide is to revise regularly in fairly small doses. This requires planning, so you will need to be organised.

ASSESSMENT

During the course of the session, you will have to complete three unit assessments, one on each area of the syllabus. The three areas are Expressions and Formulae, Relationships and Numeracy. A calculator is allowed for each of these assessments. You must pass all three assessments. You will be given the opportunity to re-sit these assessments if required; however, it is likely that you will only be given *one* opportunity to re-sit.

The final key part of the assessment will be the added value unit test. This will consist of two question papers. Paper 1 will be a non-calculator paper, while in Paper 2 you will be allowed to use a calculator. All the assessments will take place in your school or college and will be prepared and marked internally.

EQUIPMENT

A scientific calculator is recommended for this course. There will be several keys on your calculator that you will use for the first time such as those for trigonometry. It is important that your calculator is in the proper mode. At several points in the guide there will be reference to important keys and functions on your calculator. You should also be equipped with a pencil, a sharpener, a ruler, an eraser, a protractor and a pair of compasses. You will have to measure lengths and angles in some topics and draw graphs such as a pie chart, so be prepared. It is possible that your assessments may appear in a write-on paper where you have to complete a graph. It you are completing a graph or diagram in an assessment, always use a pencil so that you can erase any mistakes without making a mess.

FORMULAE LIST

In your assessments, you will have access to some important formulae.
Any formulae not on this list will have to be memorised.

Circumference of a circle: $C = \pi d$

Area of a circle: $A = \pi r^2$

Curved surface area of a cylinder: $A = 2\pi rh$

Volume of a cylinder: $V = \pi r^2 h$

Volume of a prism: $V = Ah$

Theorem of Pythagoras:

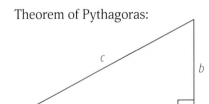

$$a^2 + b^2 = c^2$$

Trigonometric ratios in a right-angled triangle:

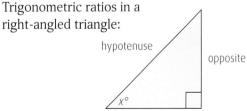

$$\tan x° = \frac{\text{opposite}}{\text{adjacent}}$$

$$\sin x° = \frac{\text{opposite}}{\text{hypotenuse}}$$

$$\cos x° = \frac{\text{adjacent}}{\text{hypotenuse}}$$

FINALLY

It is essential that regular practice forms part of your studying process. Regular practice will undoubtedly help you to improve your skills as well as giving you confidence that you are heading in the right direction.

The SQA website may help with your studying as there will be information about the course as well as practice material. Log onto SQA, choose mathematics from the subject list and search for National 4 Mathematics.

Best wishes with all your studying this session.

EXPRESSIONS AND FORMULAE

MULTIPLYING OUT BRACKETS AND FACTORISATION

WRITING IN SHORTER FORM

In this section, we will focus on the use of algebra in mathematics. We shall use algebraic skills to manipulate expressions and work with formulae.

An **expression** is a term such as $2x$ or a collection of terms such as $2x + 3y$. The letters in an expression are known as variables as they may take different numerical values.

Expressions should be written in shorter form as illustrated below.

$2 \times x$ should be written as $2x$ $3 \times y$ should be written as $3y$

$1 \times x$ should be written as x $x \times x$ should be written as x^2

$y \times x$ should be written as xy $x \div y$ should be written as $\frac{x}{y}$

DON'T FORGET

Multiplying out brackets is a fairly simple process, but its simplicity can lead to students making careless errors. For example, a common wrong answer when students are asked to multiply out $7(b + 3)$ is $7b + 3$. The student has forgotten to multiply 7×3. Therefore be careful!

EXAMPLE
Write $4y \times 3$ in shorter form.

SOLUTION
$4y \times 3 = 3 \times 4y = 12y$

MULTIPLYING OUT BRACKETS

Often expressions will contain terms which are grouped inside brackets. It is vital that you know how to multiply out (or expand) the brackets. To do this, consider the following example. Suppose you are asked to find the value of 6×14 without using a calculator. You might decide to split 14 into $(10 + 4)$ and then calculate 6×14 as $6 \times (10 + 4)$ leading to $(6 \times 10) + (6 \times 4) = 60 + 24 = 84$. This is illustrated in the diagram below.

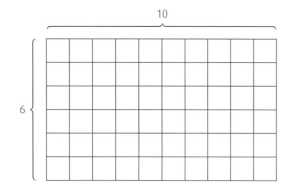
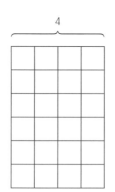

$6 \times 14 = 6 \times (10 + 4) = (6 \times 10) + (6 \times 4) = 60 + 24 = 84$

This method of finding the solution to the example illustrates the *distributive law*, which is an important law in mathematics.

The distributive law can be given in a more general form:
$a \times (b + c) = (a \times b) + (a \times c)$

The law is usually written as $a(b + c) = ab + ac$ for short.
This law can also be used in the form
$a(b - c) = ab - ac$ and can be used to multiply out brackets.

DON'T FORGET

Practice makes perfect – note that with practice, the middle steps in the following calculations can be omitted.

EXAMPLE

Multiply out the brackets:

(a) $7(b + 3)$
(b) $6(y - 4)$
(c) $3(4a + 9b)$
(d) $9(7 - 2p)$

SOLUTION

(a) $7(b + 3) = 7 \times (b + 3) = (7 \times b) + (7 \times 3) = 7b + 21$
(b) $6(y - 4) = 6 \times (y - 4) = (6 \times y) - (6 \times 4) = 6y - 24$
(c) $3(4a + 9b) = 3 \times (4a + 9b) = (3 \times 4a) + (3 \times 9b) = 12a + 27b$
(d) $9(7 - 2p) = 9 \times (7 - 2p) = (9 \times 7) - (9 \times 2p) = 63 - 18p$

FACTORISATION

Factorisation is the inverse process to multiplying out brackets. For example, we know that when we multiply out $7(b + 3)$ we get $7b + 21$. Well, if we factorise $7b + 21$, we get $7(b + 3)$. To be able to factorise, we must understand what a *factor* is. Factors are numbers which multiply together to give another number. Therefore, factors of 16 could be 2 and 8 because $2 \times 8 = 16$. The **Highest Common Factor (HCF)** of two or more numbers is the largest factor common to these numbers.

EXAMPLE

What is the highest common factor of 16 and 24?

SOLUTION

The factors of 16 are 1, 2, 4, 8 and 16.
The factors of 24 are 1, 2, 3, 4, 6, 8, 12 and 24.

The common factors of 16 and 24 are 1, 2, 4 and 8.

Hence the highest common factor of 16 and 24 is 8.

To factorise an expression, you must choose the HCF of the terms in the expression, then 'take this out' of the terms. Once the HCF has been selected, it should be fairly simple to fill in the brackets.

EXAMPLE

Factorise fully:

(a) $4x + 20$
(b) $10p - 15$
(c) $16y + 24z$
(d) $12 - 4m$

SOLUTION

(a) $4(x + 5)$
(b) $5(2p - 3)$
(c) $8(2y + 3z)$
(d) $4(3 - m)$

 **DON'T FORGET**

When you factorise an expression, you should always check your answer by multiplying out the brackets in your solution. This should lead to the original expression. It is important that you choose the *highest* common factor. For example in part (d) below, some students would choose 2 as the common factor and go on to say that $12 - 4m = 2(6 - 2m)$. While the brackets in this solution do multiply out to give $12 - 4m$, the expression has not been *fully* factorised as 2 is not the HCF of 12 and 4.

 DON'T FORGET

When factorising an expression, always start by choosing the highest common factor of the terms in the expression.

 THINGS TO DO AND THINK ABOUT

1. Write in shorter form:
    (a) $7 \times m$ (b) $p \times 4$ (c) $4 \times 2x$ (d) $3 \times q \times 5$

2. Multiply out the brackets:
    (a) $4(m + 1)$ (b) $3(x - 4)$ (c) $8(2a + 5)$ (d) $10(1 - 3z)$

3. What is the highest common factor of 24 and 36?

4. Factorise:
    (a) $5b + 30$ (b) $12a - 16$ (c) $7y + 56$ (d) $30 - 36c$

ALGEBRAIC EXPRESSIONS

SIMPLIFYING ALGEBRAIC EXPRESSIONS

You should be able to simplify algebraic expressions which have more than one variable, for example $5a + 4b + 2a$.

- In the above expression there are two variables, a and b.
- The expression has three terms.
- Each term includes its sign (+ or –), for example $5a$ means the same as $+ 5a$.
- The terms can be added in any order, for example $5a + 4b + 2a = 5a + 2a + 4b$.
- Terms such as $5a$ and $2a$ are called **like terms** and so $5a + 2a$ can be simplified leading to $7a$.
- Terms such as $7a$ and $4b$ are *not* like terms so $7a + 4b$ cannot be simplified.

Hence if you were asked to simplify $5a + 4b + 2a$, the solution would be $7a + 4b$.

It will help when you are simplifying expressions if you remember that a term like a means the same as $1a$, while a term like $-a$ means the same as $-1a$. A term such as $0a$ is the same as 0.

$2a + 3p$

EXAMPLE

Simplify:

(a) $4x + 5x - x$
(b) $10p + 2q - 7p$
(c) $a + 6b + 5a - 4b$
(d) $6x + 5y + 3x - y$

SOLUTION

(a) $4x + 5x - x = 4x + 5x - 1x = 8x$
(b) $10p + 2q - 7p = 10p - 7p + 2q = 3p + 2q$
(c) $a + 6b + 5a - 4b = 1a + 5a + 6b - 4b = 6a + 2b$
(d) $6x + 5y + 3x - y = 6x + 3x + 5y - 1y = 9x + 4y$

JUST A WEE NOTE

With practice, the middle step can be omitted.

Some examples may require you to multiply out brackets before simplifying.

EXAMPLE

Expand the brackets and simplify:

(a) $4(x + 5) + 3x$
(b) $7(x + 3y) + 2(x - 6y)$

SOLUTION

(a) $4(x + 5) + 3x = 4x + 20 + 3x = 4x + 3x + 20 = 7x + 20$
(b) $7(x + 3y) + 2(x - 6y) = 7x + 21y + 2x - 12y = 7x + 2x + 21y - 12y = 9x + 9y$

EVALUATING ALGEBRAIC EXPRESSIONS

If we are given numerical values for the variable(s) in an expression, we will be able to evaluate the expression.

EXAMPLE

If $c = 6$ and $d = 4$, evaluate:

(a) $5c$ (b) d^2 (c) $\frac{c}{d}$

SOLUTION

(a) $5c = 5 \times 6 = 30$ (b) $d^2 = 4 \times 4 = 16$ (c) $\frac{c}{d} = \frac{6}{4} = 6 \div 4 = 1\cdot5$

Order of Operations

Some expressions involve several operations – addition, subtraction, multiplication and division, as well as brackets, powers of numbers and roots. With these more complicated expressions the operations must be carried out in the correct order.

You should do things in brackets first before dealing with any orders, that is powers or roots, and then you should multiply or divide before you add or subtract.

DON'T FORGET

The correct order of operations is essential when carrying out calculations and using formulae. A useful memory aid is the mnemonic BODMAS:

B	Brackets first
O	Orders (Powers, Roots)
DM	Division and Multiplication
AS	Addition and Subtraction

EXAMPLE

If $p = 5$ and $q = 3$, evaluate (without a calculator):

(a) $6 + 7p$
(b) $3p^2$
(c) $4(p - q)$
(d) $p^2 + q^2$

SOLUTION

(a) $6 + 7p = 6 + 7 \times 5 = 6 + 35 = 41$ (multiplication before addition)
(b) $3p^2 = 3 \times 5^2 = 3 \times 25 = 75$ (powers before multiplication)
(c) $4(p - q) = 4 \times (5 - 3) = 4 \times 2 = 8$ (brackets before multiplication)
(d) $p^2 + q^2 = 5^2 + 3^2 = 25 + 9 = 34$ (powers before addition)

Check all the examples carefully to ensure you follow the correct order of operations. Make sure you follow the powers in parts (b) and (d), for example 5^2 means '5 to the power 2' or '5 squared' and equals $5 \times 5 = 25$.

COURSE IDEA

Many calculators are programmed to carry out calculations in the correct order automatically. Try the above calculations on your own calculator, working from left to right and see if you get the correct solution.

You may have to use some different keys on your calculator such as brackets and the squaring key (x^2). Find out more about these keys.

THINGS TO DO AND THINK ABOUT

1. Simplify:
   (a) $7y - y + 5y$ (b) $5p + 4q + 3p - 2q$

2. Expand the brackets and simplify:
   (a) $4(3x + 5) + 2x$ (b) $4(x + 3) + 3(x - 1)$

3. If $a = 4$ and $b = 7$, evaluate:
   (a) $5a + 2$ (b) $6a - 2b$ (c) $a^2 + b^2$ (d) $\frac{3(a + b)}{11}$

FORMULAE

WHAT IS A FORMULA?

In mathematics, a formula is a way of working out something when you know something else. For example, you can use a formula to work out the area of a rectangle when you know the length and breadth of the rectangle. If you know a formula, you can use it repeatedly, even though the numerical values might change. In other words, you can find the area of any rectangle for different lengths and breadths.

Another example might arise if you know the temperature in degrees Celsius and want to work out the temperature in degrees Fahrenheit. This can be done if you know the correct formula for converting degrees Celsius to degrees Fahrenheit.

Formulae (the plural of formula) can be given in words or symbols. You should be familiar with the formula for finding the area of a rectangle, referred to above:

Area of a rectangle = length × breadth

This formula, given above in words, can be written in symbols, namely $A = lb$.

When you are using a formula, you will have to replace the variables, length (l) and breadth (b) in the above example, with numerical values and do a calculation. The ideas on evaluating an expression from the previous section will be used to evaluate formulae.

FORMULAE IN WORDS

Although it is usual to state formulae in symbols, many formulae are just as simply explained in words.

EXAMPLE

A company hires out kilts for weddings and social events.
The formula for the cost of hiring a kilt is given in words:
Cost of hiring a kilt is £45 for 4 nights plus £10 for every extra night.
Calculate the cost of hiring a kilt for 7 nights.

SOLUTION

Cost = £45 (for 4 nights) + 3 × £10 (for 7 − 4 = 3 extra nights) = £45 + £30 = £75

It is simpler to give this formula in words than to write it in symbols, however in many cases it is much more convenient to give formulae in symbols.

FORMULAE IN SYMBOLS

The formula $A = lb$ is an example of a formula written in symbols.
The symbols in a formula could be letters (variables) or numbers such as π which is used in many calculations involving circles. Other formulae could contain mathematical symbols such as the square root sign ($\sqrt{\ }$).

When you are using a formula, it is important that you set out your working properly. You should start by writing down the formula, then replace the variable(s) by the appropriate value(s) and then carry out the calculations.

EXAMPLE

A rectangle is 15 centimetres long and 8 centimetres broad. Calculate its area.

SOLUTION

$A = lb = 15 \times 8 = 120$

Hence the area of the rectangle is $120\,\text{cm}^2$

EXAMPLE

The formula for converting degrees Celsius to degrees Fahrenheit is

$F = 1{\cdot}8C + 32$

where F is the temperature in degrees Fahrenheit and C is the temperature in degrees Celsius.

Convert 20° Celsius to degrees Fahrenheit.

SOLUTION

$F = 1{\cdot}8C + 32 = 1{\cdot}8 \times 20 + 32 = 36 + 32 = 68$

Hence the solution is 68° Fahrenheit.

EXAMPLE

John works in a factory making shirts.

His weekly salary is calculated using the formula

$S = 5{\cdot}5H + 1{\cdot}2N$

where S is his salary in pounds, H is the number of hours he works each week and N is the number of shirts he makes during the week.

One week, John works 40 hours and makes 52 shirts.

Calculate his salary for that week.

SOLUTION

$S = 5{\cdot}5H + 1{\cdot}2N = 5{\cdot}5 \times 40 + 1{\cdot}2 \times 52 = 220 + 62{\cdot}4 = 282{\cdot}4$

Hence John's salary is £282·40

WORKING BACK

The formula for converting degrees Celsius to degrees Fahrenheit is

$F = 1{\cdot}8C + 32$

Convert 50° Fahrenheit to degrees Celsius.

When given a formula, it is important to be able to work back, so study this example carefully. Note that a different approach on how to do this is shown in the section on Changing the subject of a formula on page 42.

SOLUTION

In this case, we have to work back from the formula for converting from Celsius to Fahrenheit. To use the original formula, you had to multiply by 1·8 then add 32. For this reverse calculation, you have to 'undo' these operations by subtracting 32, then dividing by 1·8.

Hence $C = (50 - 32) \div 1{\cdot}8 = 18 \div 1{\cdot}8 = 10$

Hence the solution is 10° Celsius.

Of course there is a related formula for converting from degrees Fahrenheit to degrees Celsius and we shall investigate this later.

THINGS TO DO AND THINK ABOUT

The perimeter of the shape in the diagram can be found using the formula

$P = 4a + 2b$

Calculate P when $a = 4{\cdot}5$ and $b = 7{\cdot}2$

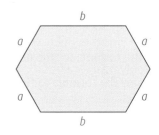

NUMBER PATTERNS

SEQUENCES

In this section, we shall consider number patterns or sequences. In mathematics, a sequence is a set of numbers that are in order, for example 4, 6, 8, 10 and so on. The numbers in a sequence are called terms. In the above sequence, the 1st term is 4, the 2nd term 6, etc. Sequences have rules for finding the terms. In this sequence the rule is 'start at 4 and add 2'.

EXAMPLE

See if you can write down the next three terms in the following sequences:

(a) 1, 3, 5, 7, ...
(b) 1, 4, 9, 16, ...
(c) 1, 3, 6, 10, 15, ...
(d) 2, 3, 5, 7, 11, 13, ...

SOLUTION

(a) 9, 11, 13
(b) 25, 36, 49
(c) 21, 28, 36
(d) 17, 19, 23

JUST A WEE NOTE

The above sequences all have names. The sequence in part (a) is odd numbers, part (b) is square numbers, part (c) is triangular numbers and part (d) is prime numbers. Investigate why square and triangular numbers are so called. Remember that **prime numbers** are numbers that are only divisible by 1 and themselves.

CONSTRUCTING A FORMULA

When we have a simple sequence formed by adding a constant amount each time, for example 4, 10, 16, 22, 28 and so on, we can construct a formula for finding *any* term in the sequence. This will be very useful if you wish to find say the 50th term of the sequence. Note that the rule for finding the next term in the sequence (add 6) is *not* a formula for finding the terms as the 50th term is clearly not going to be 56. In fact the formula for finding the terms of the sequence is a 'two-step' formula and can be found as follows:

As the sequence goes up by 6 each time, the *first step* in the formula will be '6 × n' where n is the term number. We can try this out:

n	term	first step
1	4	6 × 1 = 6
2	10	6 × 2 = 12
3	16	6 × 3 = 18

We can see that the results of the first step give a result which is too large by 2, therefore the second step must be '– 2' and the formula for this sequence will be 6 × n – 2 or $6n - 2$ for short. We can check this formula:

n	term	formula
1	4	6 × 1 – 2 = 4
2	10	6 × 2 – 2 = 10
3	16	6 × 3 – 2 = 16

We say that the formula for the nth term of the sequence is $6n - 2$.

We can now use the formula to find that the 50th term of the sequence must be 6 × 50 – 2 = 298.

EXAMPLE

Ann is drawing crosses using squares.

 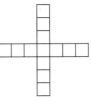

Cross Number 1   Cross Number 2   Cross Number 3

(a) Complete the table below.

Cross number (c)	1	2	3	4		10
Number of squares (s)	5	9				

(b) Write down a formula for calculating the number of squares (s) when you know the cross number (c).

(c) Ann uses 97 squares to draw a cross. What is the cross number?

SOLUTION

(a) By counting 3 → **13**, then by adding on 4 each time
4 → **17**, 5 → 21, 6 → 25, 7 → 29, 8 → 33, 9 → 37, 10 → **41**.

(b) As we are adding on 4 each time, first step in formula is × 4, then we must add 1 for the second step so formula is $s = 4c + 1$. The formula should be written in this form.

(c) Substitute $s = 97$ in the formula, leading to $4c + 1 = 97$, then work back by subtracting 1 and then dividing by 4. The solution is then $c = (97 - 1) \div 4 = 96 \div 4 = 24$. You should check your solution.

DON'T FORGET

This type of example is very important and must be practised until you gain confidence.

COURSE IDEA

Look at the sequence 1, 1, 2, 3, 5, 8, 13 and so on. What are the next three terms? This sequence is called the *Fibonacci sequence*, named after an Italian mathematician who lived around 1200. This sequence is found repeatedly in art and nature linking such varied things as a pineapple, the keys on a piano, the paintings of Leonardo da Vinci, the family tree of a bee and the petals on a buttercup. It is rewarding to investigate this sequence.

Did you get the next three terms? They are found by adding the previous two terms each time leading to 21, 34 and 55.

THINGS TO DO AND THINK ABOUT

A gardener is erecting a fence round his garden. The fence comes in sections which use lengths of wood as shown in the diagram.

 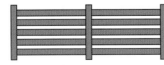 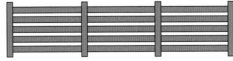

1 section   2 sections   3 sections

(a) Complete the table below.

Number of sections (s)		1	2	3	4		11
Number of lengths of wood (w)		7	13				

(b) Write down a formula for calculating the number of lengths of wood (w) when you know the number of sections (s).

(c) The gardener needs 139 lengths of wood to complete the fence. How many sections are there in the fence?

GRADIENT

The **gradient**, or slope, of a straight line is a measurement which tells us how steep it is.

The gradient of a straight line can be stated in the following formula:

$$\text{Gradient} = \frac{\text{Vertical height}}{\text{Horizontal distance}}$$

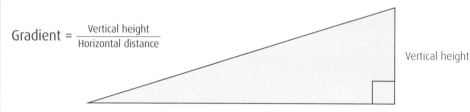

Vertical height

Horizontal distance

GRADIENT EXAMPLES

EXAMPLE

A sketch of a ramp is shown below.

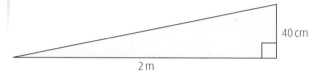

40 cm

2 m

Find the gradient of the ramp. Express your answer as a fraction in its simplest form.

SOLUTION

The measurements must have the same units, so convert metres to centimetres, changing 2 m to 200 cm.

$$\text{Gradient} = \frac{\text{Vertical height}}{\text{Horizontal distance}} = \frac{40}{200} = \frac{1}{5}$$

COURSE IDEA

Ramps are used in many real-life situations, for example, to allow wheelchair access to buildings, to load large items onto trucks and vans and even in skateboard parks. Due to health and safety considerations there are strict regulations concerning the gradient of such ramps. Investigate this.

EXAMPLE

Safety regulations state that the maximum gradient on a railway network should be 0·027.

Measurements are taken on a particular slope on the network.

 15 m

600 m

Does this slope meet the safety regulations?

Give a reason for your answer.

SOLUTION

$$\text{Gradient} = \frac{\text{Vertical height}}{\text{Horizontal distance}} = \frac{15}{600} = 15 \div 600 = 0\cdot025$$

Yes, the slope does meet the regulations because 0·025 < 0·027.

COORDINATES

When we consider gradients on a *coordinate grid*, for example finding the gradient of the straight line joining two given points, some special rules apply.

- Parallel lines have the same gradient.
- Lines which slope upwards from left to right, are said to have a *positive* gradient.
- Lines which slope downwards from left to right, are said to have a *negative* gradient.
- Lines which are horizontal, are said to have a *zero* gradient.
- Lines which are vertical, are said to have an *undefined* gradient.

These results for gradients are summarised below:

Positive	Negative	Zero	Undefined
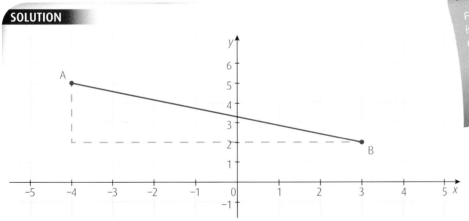			

When you are asked to find the gradient of the straight line joining two given points, you should plot the points, join them, form a right-angled triangle (if possible) and calculate the gradient using the formula given at the start of this section. You must also remember the rules given above.

EXAMPLE

Find the gradient of the straight line joining the points A (−4, 5) and B (3, 2).

JUST A WEE NOTE

For line AB, the vertical height is −3 since the height goes down by 3 and the horizontal distance is 7 as the distance goes forward by 7. Note that $\frac{-3}{7}$ would normally be written as $-\frac{3}{7}$.

SOLUTION

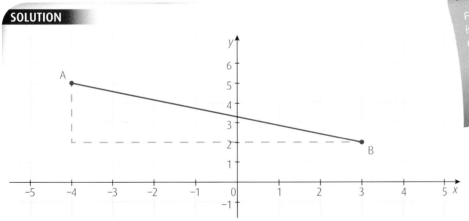

$$\text{Gradient of AB} = \frac{\text{Vertical height}}{\text{Horizontal distance}} = \frac{-3}{7} = -\frac{3}{7}$$

DON'T FORGET

If you are calculating gradients which involve coordinates, while you can still use the same formula, watch out for lines which may have a negative gradient due to the way they slope.

 THINGS TO DO AND THINK ABOUT

1. According to regulations, the gradient on a cycle track should be less than 0·05. A slope on a cycle track is shown below.

12 m
180 m

   Does this slope meet the safety regulations? Give a reason for your answer.

2. Find the gradient of the straight line joining the points A (3, 1) and B (9, 4).

THE CIRCUMFERENCE OF A CIRCLE

In mathematics, the *perimeter* of a shape is the distance around the outside of the shape. However, when we talk about the perimeter of a circle, we have a special name, that is the **circumference** of the circle.

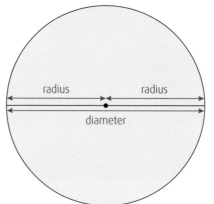

DON'T FORGET

You must know the meaning of the words 'radius' and 'diameter'. The diameter is always double the radius.

FORMULA FOR FINDING THE CIRCUMFERENCE OF A CIRCLE

COURSE IDEA

Use a measuring tape to compare the circumference and diameter of various circular objects, for example tins and cylinders. You should find that for every circle, the circumference is just over 3 × the diameter. This relationship has fascinated mathematicians for centuries. There is even a reference to it in the Bible, First Book of Kings, chapter 7, verse 23, which indicates that the circumference of a circle is 3 × the diameter.

In fact, a very accurate value has been found for this number. The number has a special symbol, π, and is referred to as pi. Press the π key on your calculator

and the following value should appear.

$$3{\cdot}141592654$$

The value of π continues forever, so we usually say that $\pi = 3{\cdot}14$ approximately. However, you should always use the π key on you calculator when carrying out circle calculations as it gives a more accurate answer. We now have a formula for finding the circumference of a circle:

Circumference of a circle = π × diameter

or $C = \pi d$ for short.

EXAMPLE

Find the circumference of a circle of diameter 15 centimetres. Give your answer to the nearest centimetre.

SOLUTION

$C = \pi d = \pi \times 15 = 47{\cdot}1238898$

Hence circumference = 47 cm
(to the nearest centimetre)

JUST A WEE NOTE

If you are unsure about rounding, go to the Rounding section on page 60.

EXAMPLE

Find the circumference of a circle of radius 13 centimetres. Give your answer to the nearest centimetre.

SOLUTION

First find the diameter, that is $d = 2 \times 13 = 26$

$C = \pi d = \pi \times 26 = 81{\cdot}68140899$

Hence circumference = 82 cm
(to the nearest centimetre)

JUST A WEE NOTE

If you are given the radius of a circle and asked to find the circumference, you should double the radius first to get the diameter. However, it should be pointed out that the formula for the circumference can also be given in the form $C = 2\pi r$. In this form the radius can be substituted directly into the formula.

PERIMETER

You may be asked to find the perimeter of a shape, part of which involves a calculation of the circumference of a circle.

The shooting area on a netball court is a semi-circle of radius 4·9 metres.

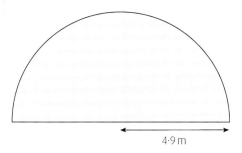

4·9 m

Calculate the perimeter of the shooting area.

Give your answer correct to the nearest metre.

Note that a semi-circle is exactly half of a circle. The perimeter of the shooting area consists of a semi-circular edge and a straight edge.

Straight edge (diameter) = 2 × 4·9 = 9·8

Semi-circular edge = $\frac{1}{2}\pi d$ = 0·5 × π × 9·8 = 15·393804

Now add edges leading to 9·8 + 15·393804 = 25·193804

Hence perimeter of shooting area = 25 m (to the nearest metre)

WORKING BACK

Finally in this section, we shall look at how you can work back from the circumference of a circle to find the diameter or radius. When we find the circumference of a circle, we use the formula $C = \pi d$ and this involves multiplication. Therefore when we are working back, we must use the inverse operation, namely division.

The circumference of a circle is 90 centimetres. Calculate the diameter of the circle.

Give your answer to the nearest centimetre.

To find the circumference, you have to multiply the diameter by π. For the reverse calculation, you have to 'undo' this operation by dividing the circumference by π.

$d = 90 \div \pi = 28\cdot64788976$

Hence diameter = 29 cm (to the nearest cm)

JUST A WEE NOTE

If you are told the circumference and asked to find the radius, divide the circumference by π to get the diameter and then divide the diameter by 2 to get the radius.

THINGS TO DO AND THINK ABOUT

Give your answer to the nearest centimetre in each question.

1. Find the circumference of a tyre of diameter 63 centimetres.
2. Find the circumference of a circle of radius 11 centimetres.
3. Find the diameter of a circle whose circumference is 113 centimetres.
4. The minute hand on a wall clock is 20 centimetres long.
   What distance will the tip of the minute hand travel in an hour?

THE AREA OF A CIRCLE

REMINDERS

Remember the main metric units of length:

1 cm = 10 mm   1 m = 100 cm   1 km = 1000 m

We shall now look at area, a measure of the size of any surface. Remember that area is measured in square units.

The rectangle in this diagram has an area of 10 square units.

Area can be measured in square millimetres (mm²) for small areas such as a postage stamp, square centimetres (cm²) for areas such as a page of this book, square metres (m²) for areas such as a football pitch and square kilometres (km²) for larger areas such as countries.

Another unit of area is a *hectare*. 1 hectare = 10 000 m² and is used for measuring land areas such as a farm.

CALCULATING AREAS

You should already know how to calculate the area of basic shapes such as a rectangle, a square and a triangle. The formulae are given below:

Area of a rectangle = length × breadth   or $A = lb$

Area of a square = length × length   or $A = l^2$

Area of a triangle = $\frac{1}{2}$ × base × height   or $A = \frac{1}{2}bh$

EXAMPLE

Calculate the area of the triangle shown in the diagram.

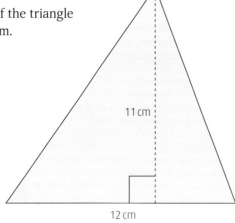

11 cm

12 cm

SOLUTION

$A = \frac{1}{2}bh = 0{\cdot}5 \times 12 \times 11 = 66$

Hence area of triangle = 66 cm²

CIRCLES

Mathematicians have found a formula for calculating the area of a circle. As with the circumference, the formula involves π. The formula is given below.

Area of a circle: $A = \pi r^2$

You can use this formula to find the area of any circle as long as you know the radius.

In each of the following examples, round your answer to the nearest square centimetre.

EXAMPLE

An Indian dholak drum is shown.

The top of the drum has a radius of 11·5 centimetres.

Calculate the area of the top of the drum.

SOLUTION

$A = \pi r^2 = \pi \times 11\cdot5^2 = 415\cdot4756284$

Hence area of top of drum = 415 cm² (to the nearest cm²)

 DON'T FORGET

It is simple to carry out this calculation using the x^2 key on your calculator. Try using this key on the above calculation and practise using it until you are confident. If you do not have this key, calculate $\pi \times 11\cdot5^2$ by doing $\pi \times 11\cdot5 \times 11\cdot5$.

EXAMPLE

A circular jigsaw puzzle has a diameter of 50 centimetres. Calculate its area.

SOLUTION

First find the radius, that is $r = 50 \div 2 = 25$.

$A = \pi r^2 = \pi \times 25^2 = 1963\cdot495408$

Hence area of jigsaw puzzle = 1963 cm² (to the nearest cm²)

COMPOSITE SHAPES

A **composite** shape is a shape made up of two or more shapes. The area of a composite shape can be found by adding (or subtracting) other areas.

EXAMPLE

The badge for a school blazer is being designed. The badge is in the shape of a square and a semi-circle. The dimensions of the badge are shown below.

6 cm

Calculate the area of the badge.

SOLUTION

To find the area of the badge, add the area of a square and the area of a semi-circle.

Area of square: $A = l^2 = 6^2 = 36$

Radius of semi-circle = $6 \div 2 = 3$

Area of semi-circle: $A = \frac{1}{2}\pi r^2 = 0\cdot5 \times \pi \times 3^2 = 14\cdot13716694$

Area of badge = $36 + 14\cdot13716694 = 50\cdot13716694$

Hence area of badge = 50 cm² (to the nearest cm²)

BEWARE: When carrying out calculations involving circles, it is common for students to get the formulae for area and circumference mixed up. So be careful!

 THINGS TO DO AND THINK ABOUT

1. The diameter of the centre circle on a football pitch is 18·3 metres.
   (a) Calculate the circumference of the centre circle.
   (b) Calculate the area of the centre circle.

2. A church window is in the shape of a rectangle with a semi-circular top.
   Calculate the area of the church window in square metres.
   Give your answer correct to one decimal place.

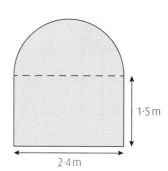

1·5 m

2·4 m

THE AREA OF QUADRILATERALS

SPECIAL QUADRILATERALS AND THEIR PROPERTIES

A **quadrilateral** is a shape with four sides which are all straight lines. Some quadrilaterals have special names. We shall consider different types of quadrilateral and look at their properties.

A rectangle has opposite sides equal and parallel. It has four right angles. Its diagonals are equal. The diagonals bisect one another, that is, they cut each other in half. It has two axes of symmetry (not the diagonals).

A square has all four sides equal. Its opposite sides are parallel. It has four right angles. Its diagonals are equal. The diagonals bisect one another at right angles. The diagonals bisect the four right angles. It has four axes of symmetry.

A parallelogram has opposite sides equal and parallel. Its opposite angles are equal. The diagonals bisect one another. It has no axes of symmetry.

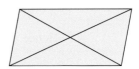

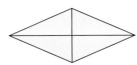
A rhombus has all four sides equal. Its opposite angles are equal. The diagonals bisect one another at right angles. The diagonals bisect the four angles of the rhombus. It has two axes of symmetry (the diagonals).

A kite has two pairs of adjacent sides equal. The diagonals cut one another at right angles. There is one axis of symmetry, which is one of the diagonals. This diagonal bisects the other diagonal and the two angles at its ends.

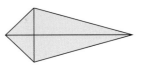

A trapezium has one pair of parallel sides. In general, it has no axis of symmetry although *some* trapeziums are symmetrical.

In *all* quadrilaterals, the four angles always add up to 360°.

QUADRILATERALS AND THEIR AREAS

We already know how to find the area of a rectangle and square. We shall now look at how to calculate the areas of the other special quadrilaterals. To do this, we shall think of them as being composite shapes made up of rectangles and/or triangles. Study the following example.

DON'T FORGET

It is important that you are aware of all the different types of quadrilateral. Many of the properties are obvious but make sure that you understand all the different terms involved.

EXAMPLE

The top of a table is in the shape of a trapezium.

The trapezium is symmetrical with measurements shown below.

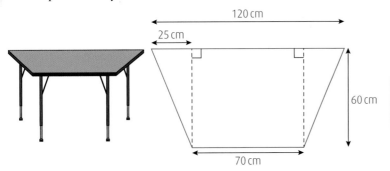

Find the area of the top of the table.

SOLUTION

Area of rectangle: $A = lb = 70 \times 60 = 4200$

Area of one right-angled triangle:
$A = \frac{1}{2}bh = 0{\cdot}5 \times 25 \times 60 = 750$

Hence area of table top
$= (4200 + 750 + 750)\,\text{cm}^2 = 5700\,\text{cm}^2$

USING SUBTRACTION TO FIND AREA

EXAMPLE

The shape below is sometimes called a V-kite.
Calculate its area.

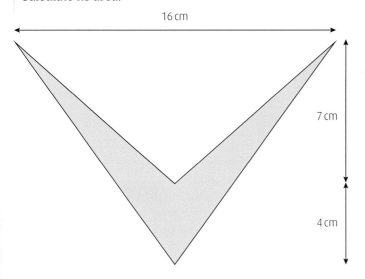

SOLUTION

We can find the area of the V-kite by subtracting the area of the smaller triangle from the area of the larger triangle. The larger triangle has base 16 centimetres and height (7 + 4) = 11 centimetres. The smaller triangle has base 16 centimetres and height 7 centimetres. Make sure you understand these steps.

Area of larger triangle = $\frac{1}{2}bh$ = 0·5 × 16 × 11 = 88

Area of smaller triangle = $\frac{1}{2}bh$ = 0·5 × 16 × 7 = 56

Hence area of V-kite = (88 – 56) cm² = 32 cm²

COURSE IDEA

We have used the idea of composite shapes to calculate the areas of some special quadrilaterals. However there are formulae for these shapes, namely $A = bh$ for the parallelogram, $A = \frac{1}{2}d_1 d_2$ for the kite and rhombus and $A = \frac{1}{2}(a + b)h$ for the trapezium.

It is very efficient to use these formulae when required. There is a disadvantage however. You will have to memorise the formulae and learn what the various symbols represent. However, it is definitely worthwhile investigating these methods.

 THINGS TO DO AND THINK ABOUT

The top of a set of bathroom scales is in the shape of a trapezium.
The trapezium is symmetrical with measurements as shown below.

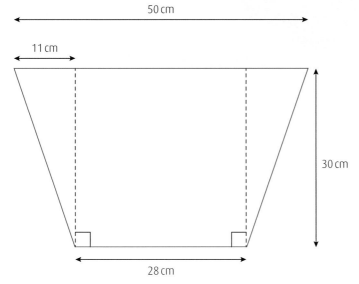

Calculate the area of the top of the set of bathroom scales.

SOLID SHAPES AND THEIR NETS

THE PRISM

You are already familiar with some 3-dimensional shapes or solids.

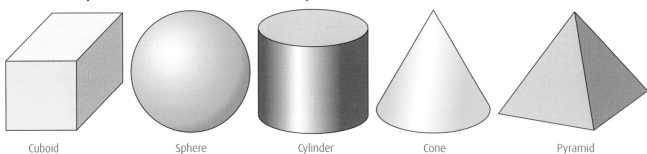

| Cuboid | Sphere | Cylinder | Cone | Pyramid |

The cuboid is an example of a **prism**. A prism is a solid shape which has opposite ends that are parallel and congruent. Congruent simply means that the two ends are equal in size and shape. The ends are referred to as the bases of the prism. If you were to make a cut through a prism parallel to the base, the cross-section would be identical to the base.

Prisms and pyramids take their names from their base.
A triangular prism (see picture) is a very familiar type of prism.

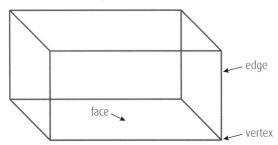

FACES, VERTICES AND EDGES

We talk about the faces, vertices and edges of solid shapes. A face is a surface of the shape, for example a die (singular of dice) has six faces. A vertex (plural: vertices) is a corner and an edge is a straight line joining one vertex to another.

Consider a cuboid.

A cuboid has six faces, eight vertices and 12 edges.

Make sure you follow this.

EXAMPLE

How many faces, vertices and edges does each of the following solid shapes have?

(a) cube
(b) triangular prism
(c) square-based pyramid

SOLUTION

Shape	Faces	Vertices	Edges
Cube	6	8	12
Triangular prism	5	6	9
Square-based pyramid	5	5	8

COURSE IDEA

A famous Swiss mathematician called Leonhard Euler lived from 1707 to 1783. A formula, which connects the number of faces, vertices and edges of solid shapes with straight edges is named after him. It is called *Euler's formula* and states that $F + V - E = 2$. Check that for a cube $6 + 8 - 12 = 2$. Find out more about Euler's formula and check it works for parts (b) and (c).

SKELETON MODELS

A skeleton model could be made with straws cut to the correct length and pipe cleaners. It gives a 'see-through' version of a solid shape. To make a skeleton model of a cube, you would need 12 equal straws and eight pieces of pipe cleaner.

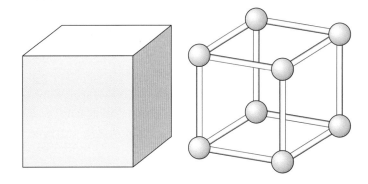

EXAMPLE

Find the total length of the straws required to make a skeleton model of a cuboid 10 centimetres long, 8 centimetres broad and 6 centimetres high.

SOLUTION

You will need four straws for each of the three dimensions.

Total length = $(4 \times 10) + (4 \times 8) + (4 \times 6) = 40 + 32 + 24 = 96$

Hence the total length of straws required is 96 cm.

NETS OF SOLIDS

The **net** of a solid shape is a flat, two-dimensional shape that can be cut out and folded up to make the solid shape. Here are two examples of the net of a cube. There are other possible nets for a cube.

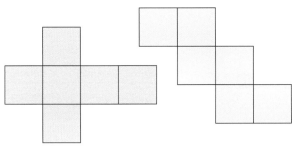

Note that both nets have the six faces required for a cube. If you are planning to make a cube from its net, you will need to add flaps for gluing it together. We shall look at this later.

COURSE IDEA

If you wanted to make a dunce's hat, you would have to start with the net of a cone. This is difficult because a cone has a curved face. Go online and investigate the nets of solid shapes with curved faces such as a cone, a cylinder and a sphere.

THINGS TO DO AND THINK ABOUT

1. A solid shape is shown below.
   How many faces, vertices and edges does this shape have?

2. What name would be given to the shape in question 1?

3. Calculate the total length of the straws you would need to make a skeleton model of a cube with sides of length 14 centimetres.

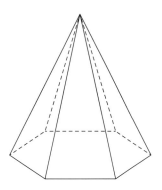

SURFACE AREA

SURFACE AREA: AN OVERVIEW

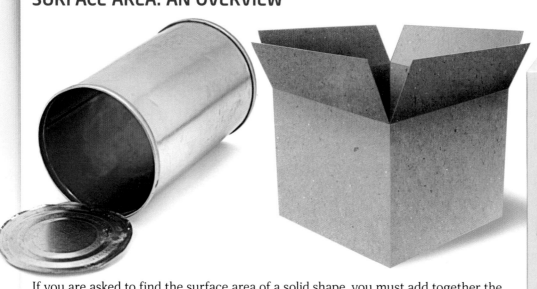

If you are asked to find the surface area of a solid shape, you must add together the area of all the faces of the solid shape. The ideas of surface area are very important when packaging and labelling items. Before companies create milk cartons, cardboard boxes, soup tins and thousands of other items found in shops around the country, someone will have designed the container, sizes will have been calculated and a net of the solid will have been prepared.

EXAMPLE

A gift box is in the shape of a cube of side 8 centimetres.

Calculate the surface area of the box.

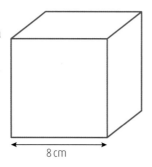

8 cm

SOLUTION

Area of each square face: $A = l^2 = 8^2 = 64$

Total surface area $= 6 \times 64 = 384$

Hence the surface area of the box is $384 \, \text{cm}^2$

EXAMPLE

Find the surface area of this cuboid.

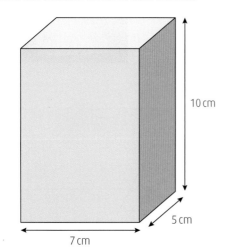

10 cm

5 cm

7 cm

SOLUTION

Front/back: $A = lb = 7 \times 10 = 70$

Right side/left side: $A = lb = 5 \times 10 = 50$

Top/bottom: $A = lb = 7 \times 5 = 35$

Total surface area $= (2 \times 70) + (2 \times 50) + (2 \times 35)$
$$= 140 + 100 + 70 = 310$$

Hence surface area $= 310 \, \text{cm}^2$

Sometimes you may be given the net of a solid shape and asked to calculate the surface area of the solid shape. If so, simply add together the areas of all the faces in the net, excluding any flaps.

EXAMPLE

The diagram below shows the net of a solid shape.

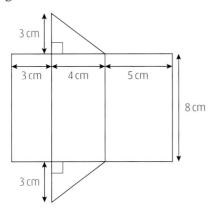

3 cm

3 cm   4 cm   5 cm

8 cm

3 cm

(a) Name the solid shape formed from this net.
(b) Calculate the surface area of this solid shape.

SOLUTION

(a) Triangular prism.
(b) Area of three rectangles = (3 × 8) + (4 × 8) + (5 × 8)
$$= 24 + 32 + 40 = 96$$
Area of two triangles $= 2 \times \frac{1}{2}bh = 2 \times 0{\cdot}5 \times 4 \times 3$
$$= 12$$

Total surface area = 96 + 12 = 108

Hence surface area = 108 cm²

THE SURFACE AREA OF A CYLINDER

A cylinder has three faces, two flat circular faces at each end and a curved surface. If you think about placing a label around the curved surface, you should realise that the label will have to be rectangular with its length equal to the circumference of the circle and its breadth equal to the height of the cylinder. In this way, we can find a formula for the *curved surface area* of a cylinder.

The net of a cylinder is illustrated below.

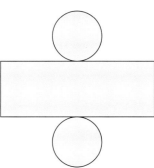

Curved surface area of a cylinder = length × breadth
$$= \text{circumference of circle} \times \text{height of cylinder}$$
$$= \pi d \times h$$
$$= 2\pi rh$$

Hence the formula for the curved surface area of a cylinder is $A = 2\pi rh$.
You will not have to memorise this formula.

EXAMPLE

A tin of soup has radius 4 centimetres and height 12 centimetres.

Calculate the curved surface area of the tin to the nearest square centimetre.

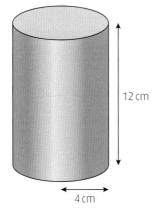

12 cm

4 cm

SOLUTION

$A = 2\pi rh = 2 \times \pi \times 4 \times 12 = 301{\cdot}5928947$

Hence the curved surface area = 302 cm² (to the nearest cm²)

✚ DON'T FORGET

Students sometimes mix up units of area and volume. Remember it is square units for area, for example, cm² and cubic units for volume, for example, cm³.

🎈 THINGS TO DO AND THINK ABOUT

1. Find the surface area of a cube of side 12 centimetres.
2. Find the surface area of a cuboid 8 centimetres long, 6 centimetres broad and 5 centimetres high.
3. Find the curved surface area of a cylinder with radius 8 centimetres and height 12 centimetres.
   Give your answer to the nearest square centimetre.
   Curved surface area of a cylinder is given by $A = 2\pi rh$.

VOLUME OF A PRISM

VOLUME

We use the term volume to measure the amount of space taken up by an object. Volume is measured in cubic units, for example cubic millimetres (mm^3), cubic centimetres (cm^3) or cubic metres (m^3). Other important measures of volume are the litre (l) and the millilitre (ml) where $1\ l = 1000$ ml or $1000\ cm^3$

You should have already used two volume formulae.

    Volume of a cuboid = length × breadth × height   or $V = lbh$ for short

    Volume of a cube = length × length × length   or $V = l^3$ for short

EXAMPLE

A fish tank is in the shape of a cuboid.

It is 60 centimetres long, 25 centimetres broad and 20 centimetres high.

Calculate its volume in litres.

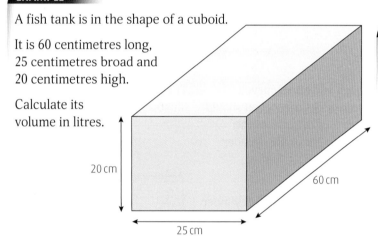

20 cm

25 cm

60 cm

SOLUTION

$V = lbh = 60 × 25 × 20 = 30\ 000$

Hence volume = $30\ 000\ cm^3$

Volume in litres = $(30\ 000 ÷ 1000)$ litres = $30\ l$

WORKING BACK

If we know the volume of a cuboid and two of the dimensions of the cuboid, we can work back to find the missing dimension. This will involve division.

EXAMPLE

The volume of a box used for chocolates is 540 cubic centimetres. If the box is 15 centimetres long and 3 centimetres high, find its breadth.

SOLUTION

$V = lbh \Rightarrow 540 = 15 × b × 3 \Rightarrow b = 540 ÷ (15 × 3)$
$= 540 ÷ 45 = 12$

Hence the box is 12 cm broad.

THE VOLUME OF A PRISM

The volume of a prism is decided by two things – the shape of the base and the height. The formula for finding the volume of all prisms is given below.

    Volume of a prism = area of base × height   or $V = Ah$ for short.

A cylinder is a special type of prism with a circular base. The formula for its volume is given below.

Volume of cylinder = $\pi r^2 h$ or $V = \pi r^2 h$ for short.

EXAMPLE

A can of fruit juice is in the shape of a cylinder. The can has height 9 centimetres and the area of the base of the can is 50 square centimetres.

Calculate the volume of the can.

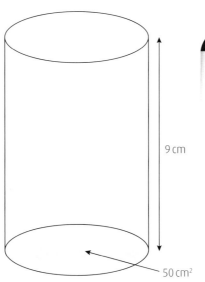

9 cm

50 cm²

SOLUTION

$V = Ah = 50 \times 9 = 450$

Hence the volume of the can is 450 cm³

 **DON'T FORGET**

When calculating the volume of a cylinder, use $V = Ah$ if you are told the height and the area of the base. Use $V = \pi r^2 h$ if you are told the height and the radius (or diameter). An example on the second formula appears in the Rounding section on page 60.

Some volume problems may cover more than one part of the syllabus. This could require some reasoning skills on your part.

EXAMPLE

Rashid is going to construct a cube from its net. He has drawn the net on a piece of cardboard 22 centimetres long and 17 centimetres broad.

The flaps for sticking the cube together are all 1 centimetre wide.

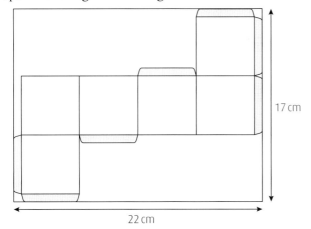

17 cm

22 cm

What is the volume of Rashid's cube?

SOLUTION

You must start by finding the length of a side of the cube.

Length of side = (22 – 2) ÷ 4 = 20 ÷ 4 = 5
or
Length of side = (17 – 2) ÷ 3 = 15 ÷ 3 = 5

Volume of cube: $V = l^3 = 5^3 = 125$

Hence the volume is 125 cm³

COURSE IDEA

To work out 5^3, you can do 5 × 5 × 5 or use the key for calculating powers on your calculator, for example, $5 \wedge 3$ or $5\, y^x\, 3$. This key varies from calculator to calculator, so find out how it works on your calculator.

THINGS TO DO AND THINK ABOUT

1. Find the volume of a tank in the shape of a cuboid measuring 12 metres by 10 metres by 5 metres.
2. A stock cube has length of side 2 centimetres. What is its volume?
3. The area of the base of a triangular prism is 22 square centimetres. Its height is 7·5 centimetres. Calculate the volume of the triangular prism.

ROTATIONAL SYMMETRY

LINE SYMMETRY

In mathematics, one type of symmetry is called **line symmetry** in which one half is the reflection of the other half. The axis of symmetry is a line where you could fold the image so that both halves match exactly.

Check the earlier section on quadrilaterals where the number of axes of symmetry of several types of quadrilateral was stated. Many other shapes have an axis of symmetry, for example an equilateral triangle has three axes of symmetry.

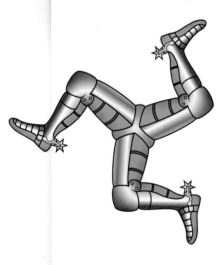

ANOTHER TYPE OF SYMMETRY

When a shape has line symmetry, if you copy it onto tracing paper and turn the paper over, you will be able to fit the shape over its outline. There is another type of symmetry where the tracing of a shape fits its outline without turning the paper over. Instead this is done by turning the tracing paper round about a point. This type of symmetry is called **rotational symmetry**. When a shape has rotational symmetry, it fits its outline as it turns (or rotates). Note that some shapes have line symmetry *and* rotational symmetry, for example a square.

ORDER OF ROTATIONAL SYMMETRY

As we turn a shape around once, that is a complete turn of 360°, the number of times a shape fits its outline is called its order of rotational symmetry. The **symbol** in the diagram above, which shows the flag of the Isle of Man, has rotational symmetry of order 3.

Draw accurate sketches of the following shapes:

(a) a square
(b) a rectangle
(c) a rhombus
(d) a parallelogram.

Trace each shape, keeping the tracing paper over the shape and rotate each shape around its centre (the point where the diagonals intersect). By counting how many times each shape fits its outline in a complete turn, write down the order of rotational symmetry for each shape.

When a shape has order of rotational symmetry 2, this is often called half-turn symmetry and means that the shape fits its outline when turned through 180°. Shapes which have half-turn symmetry look exactly the same upside down. Examples include a parallelogram and some capital letters in the alphabet, namely H, I, N, S, X and Z. Turn this page upside down to confirm this.

When a shape has order of rotational symmetry 4, this is often called quarter-turn symmetry and means that the shape fits its outline when turned through 90°. An example of this is a square.

SOLUTION

(a) 4   (b) 2   (c) 2   (d) 2

COURSE IDEA

It will help considerably to understand rotational symmetry if you use tracing paper when practising examples. If you are given a shape to rotate, trace the whole diagram, put your finger on the tracing paper over the centre of the shape and start to rotate the paper. Once you have practised this, you should be able to see how the process works and you should eventually manage without tracing paper.

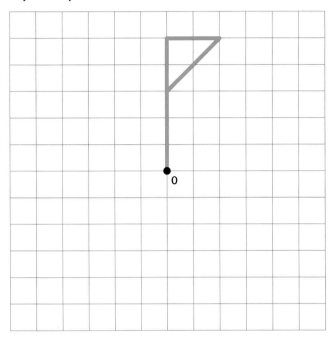

EXAMPLE

Copy the diagram below onto squared paper.

Part of the logo for an IT company is shown below.

Complete this shape so that it has rotational symmetry of order 4 about O.

SOLUTION

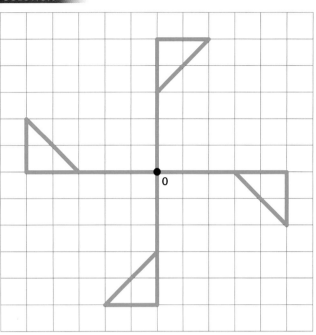

 ## THINGS TO DO AND THINK ABOUT

Copy the diagrams below onto squared paper.

1. Copy and complete this shape so that it has half-turn symmetry about O.

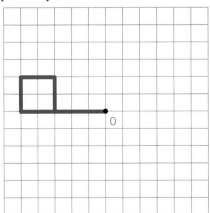

2. The Maltese Cross is a symbol of the island of Malta. The Cross has rotational symmetry of order 4. Copy and complete the cross.

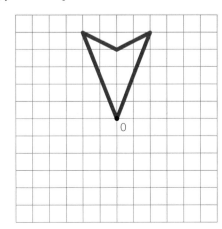

FREQUENCY TABLES

A REMINDER

You should be familiar with the idea of frequency tables and the use of tally marks in constructing them. Look at the example given below.

EXAMPLE

Jonathon rolls a die 30 times and records the following scores:

3 4 6 4 3 1 5 6 3 3 2 2 6 1 1

5 3 6 2 1 3 1 6 4 1 3 5 5 5 2

Illustrate this data in a frequency table.

SOLUTION

Score	Tally	Frequency
1	‖‖	6
2	‖‖	4
3	‖‖ ‖	7
4	‖‖	3
5	‖‖	5
6	‖‖	5

Note that it is always recommended that you add the frequencies in the completed table. This should come to 30, which agrees with how many numbers appear in the list of recorded scores.

CLASS INTERVALS

When the numbers recorded during the collection of data are widely spread, it is impractical to list all the numbers individually. In this case, numbers are grouped in **class intervals**. For example, percentages might be grouped in class intervals such as 0 – 9, 10 – 19, 20 – 29, 30 – 39, etc.

EXAMPLE

The marks of a group of students in a test marked out of 50 are listed below.

43 27 32 39 17 39 40 18 30 22 28 32 37 19 23

23 27 29 31 44 21 25 29 33 38 40 36 31 29 27

Illustrate this data in a frequency table using class intervals 15 – 19, 20 – 24, 25 – 29, etc.

SOLUTION

Marks	Tally	Frequency
15 – 19	‖‖	3
20 – 24	‖‖	4
25 – 29	‖‖ ‖‖	8
30 – 34	‖‖ ‖	6
35 – 39	‖‖	5
40 – 44	‖‖	4

DON'T FORGET

Frequency tables are fairly simple to construct but require care. Always double-check the class intervals by continuing a sequence. Also double-check the tally marks as well as adding the frequencies to make sure nothing has been omitted.

You have to continue the class intervals beyond 25 – 29. You should find the largest number in the list (44) as this tells you the final class interval. To continue the class intervals, simply continue the sequences of numbers *carefully*. Remember to add the frequencies as a check.

The results of the frequency table could be illustrated in a graph.

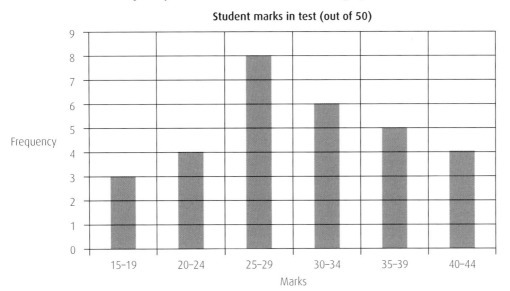

Student marks in test (out of 50)

The data is shown in the bar graph above. You should always give graphs a title, clearly label both axes and use suitable scales for both axes.

DISCRETE AND CONTINUOUS VARIABLES

Numerical data collected in surveys is variable. This means that it can take different values. There are two types of numerical data.

One type is called **discrete** data. Discrete data can only take certain values. For example, when rolling a die, you can only score 1, 2, 3, 4, 5 or 6. A bar graph is suitable for illustrating discrete data.

The other type is called **continuous** data. Continuous data can take any value in a given range. For example, the length of time taken to complete a race could be measured to fractions of a second.

COURSE IDEA

The bar graph used to show the student marks in a test would be unsuitable for illustrating continuous data as there are gaps in the horizontal scale. For example, the horizontal scale on the bar graph is labelled 15 – 19, 20 – 24, etc. so there is no place for a variable such as 19·5. Continuous data is usually illustrated in a graph known as a histogram. Find out about histograms.

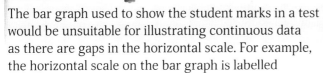

THINGS TO DO AND THINK ABOUT

1. The percentage marks of a group of students in their geography exam are listed below.

    43 57 62 39 77 59 60 58 80 42 58 62 77 59 63

    53 67 89 41 64 51 75 69 53 78 60 46 71 59 57

    Illustrate this data in a frequency table using class intervals 30 – 39, 40 – 49, 50 – 59, etc.

2. State whether the following variables are discrete or continuous.
    (a) golf scores
    (b) daily rainfall
    (c) temperature
    (d) number of matches in a box.

AVERAGES AND COMPARING DATA

THE MEAN

There are three averages commonly used in mathematics. We shall look at the **mean** first. When a report in the news mentions the word average, for example 'average house price', it is really the mean that is being discussed. The definition of the mean for a set of numerical data is given below.

$$\text{The mean} = \frac{\text{Total of all values}}{\text{Number of values}}$$

EXAMPLE

Carly records the following scores in a series of golf matches.

74 72 78 76 80 80 79 73 75 77

Calculate her mean score.

SOLUTION

$$\text{The mean} = \frac{\text{Total of all values}}{\text{Number of values}}$$

$$= \frac{74 + 72 + 78 + 76 + 80 + 80 + 79 + 73 + 75 + 77}{10}$$

$$= \frac{764}{10} = 76\cdot4$$

THE MEDIAN

Another useful average in mathematics is the **median**. The median is defined as the middle value in a set of ordered values. The median is easy to find when there is an odd number of values in a data set. For example, the median of 7, 8, 9, 10 and 11 is 9. It is not always so easy when there is an even number of values. In this case, the median lies in between two values and is found by calculating the mean of these two values.

To help, we have a formula for calculating the *position* of the median in a list of ordered data.

In a set of ordered data with n values, the position of the median is $(n + 1) \div 2$.

Therefore in the above example, where $n = 5$, the position of the median could be found by using the formula leading to $(5 + 1) \div 2 = 3$. Hence the median was the 3rd number in the ordered list.

DON'T FORGET

As they all begin with the letter M, it is easy to mix up the mean, median and mode. Find a strategy to ensure that you know exactly which is which.

EXAMPLE

Find the median for Carly's scores from above.

Order the data → 72 73 74 75 <u>76 77</u> 78 79 80 80

SOLUTION

As $n = 10$, position of median = $(10 + 1) \div 2$ = 5·5. This means the median is between the 5th and 6th values (underlined). The median is therefore the mean of 76 and 77 which is calculated as $(76 + 77) \div 2 = 153 \div 2 = 76\cdot5$. Hence the median = 76·5.

THE MODE

The third average is called the **mode**. The mode is the most frequent value in a data set.

EXAMPLE

Find the mode for Carly's scores from the above.

SOLUTION

By looking through the ordered list of Carly's scores above, we can see that the mode = 80 (the most common value in the list as it is the only value to appear twice).

WHICH AVERAGE?

We have three averages, but which is the best to use? Well, sometimes the mean is not typical of the data if there is a value which lies well away from the majority of the values.

EXAMPLE

The attendances at six premier league games on one weekend are shown below.

7520 8150 10480 57980 9060 7850

The median attendance is 8605.

(a) Calculate the mean attendance.
(b) Which of the two 'averages' – the mean or the median – is more representative of the data? You must explain your answer.

SOLUTION

(a) mean $= \dfrac{\text{Total of all values}}{\text{Number of values}}$

$$= \frac{7520 + 8150 + 10480 + 57980 + 9060 + 7850}{6}$$

$$= \frac{101040}{6} = 16840$$

(b) The median is more representative as five of the attendances are fairly close to 8605. The mean is not typical of the attendances as it is affected by one very high value.

So, although the mean is often an excellent average to use, the median is better if the mean is affected by extreme values. The mode is useful when personal views are being investigated, for example asking a group of people for their lucky number.

COMPARING DATA SETS

We can compare two data sets by considering their averages. However, it is also useful to look at how the values are spread out. We can see, for example, that the mean of the data set 7, 8 and 9 is the same as the mean of the data set 0, 8 and 16 (the mean is 8 each time), yet the data sets are very different. We therefore can use a measure of spread known as the **range**.

For any numerical data set, the range = highest value – lowest value.

For Carly's golf scores earlier, check that the range = 80 – 72 = 8.

DON'T FORGET

When comparing data sets, it is not enough to simply compare numbers and say that one is greater/smaller than another. You are expected to reach some kind of conclusion, for example whether members lost weight or not and how the results are spread out.

EXAMPLE

The weights, in kilograms, of the members of a weightwatchers club were measured.

71 84 92 80 76 67 90 80

(a) Calculate (i) the mean;
          (ii) the range.

After attending the club for 8 weeks, the members were weighed again. The mean was 72 kilograms and the range was 30 kilograms.

(b) Write two comments comparing the weights before and after attending the club.

SOLUTION

(a) (i) mean $= \dfrac{\text{Total of all values}}{\text{Number of values}}$

$$= \frac{71 + 84 + 92 + 80 + 76 + 67 + 90 + 80}{8} = \frac{640}{8} = 80 \text{ kg}$$

(ii) range = 92 – 67 = 25 kg

(b) After attending the club, members have lost weight as the mean falls from 80 kg to 72 kg. After attending the club, the weights are more spread out as the range increases from 25 to 30 kg.

THINGS TO DO AND THINK ABOUT

Find the mean, median, mode and range for the following exam marks.

40 68 42 59 63 50 48 63 58 71

PIE CHARTS

INTERPRETING A PIE CHART

Pie charts are a very attractive and clear way of representing data. In a pie chart, sectors of a circle are used to show different pieces of information.

EXAMPLE

A survey was carried out to find out what the favourite sports were in a group of 120 boys. The results are illustrated in the pie chart below.

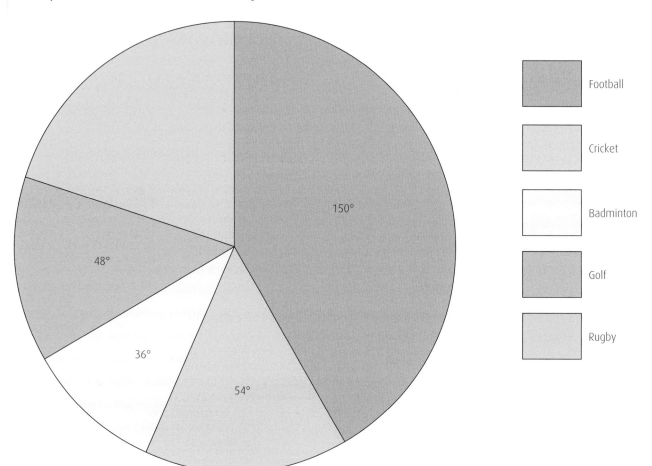

How many of the boys chose rugby?

SOLUTION

Angle sizes have been entered in every sector except rugby, so we must find the angle size for the rugby sector. To do this, remember that there are 360° in a complete turn.

Hence angle in rugby sector = 360° – (150 + 54 + 36 + 48)° = 72°

Now express 72° as a fraction of 360° and find this fraction of 120 (the number of boys in the survey).

Hence number of boys who chose rugby = $\frac{72}{360} \times 120 = 24$

If you are unsure about fractions, there is a section on fractions on page 64.

DON'T FORGET

When you are carrying out calculations involving pie charts, it is likely that 360, the number of degrees in a complete turn, will be involved.

CONSTRUCTING A PIE CHART

We shall now look at how to construct a pie chart given a set of data.

EXAMPLE

A sample of school leavers was surveyed on their destinations. The following results were obtained.

120	University/College
45	Employment
15	Training
20	Unemployed

Illustrate this data in a pie chart.

SOLUTION

Start by finding the total number of school leavers surveyed:
120 + 45 + 15 + 20 = 200

Then express each category as a fraction of the total: $\frac{120}{200}, \frac{45}{200}, \frac{15}{200}, \frac{20}{200}$

Next find each fraction of 360 in turn: $\frac{120}{200} \times 360, \frac{45}{200} \times 360, \frac{15}{200} \times 360, \frac{20}{200} \times 360$

Finally, calculate the angles for each sector: 216°, 81°, 27°, 36°
(Check these add up to 360°)

Now draw the pie chart moving clockwise from a line at 12 o'clock. Label each sector and give the pie chart a title.

DON'T FORGET

It is vital that you can use a protractor to measure angles accurately including reflex angles such as 216°. You must also be able to draw an accurate circle using a pair of compasses. If you are unsure of these important skills, seek out advice.

Destination of school leavers

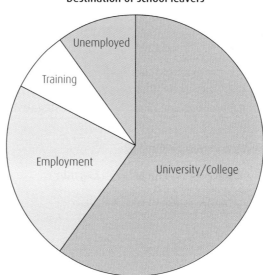

COURSE IDEA

Find out how to design a pie chart on a computer, for example, using Microsoft Excel.

 THINGS TO DO AND THINK ABOUT

100 grams of wholemeal bread contain 15 grams of protein, 45 grams of carbohydrates, 10 grams of fibre, 5 grams of fat and 25 grams of other ingredients.

This data will be illustrated on a pie chart.

What size of angle should be used for carbohydrates?

PROBABILITY

We cannot be certain what is going to happen when certain events take place, for example tossing a coin. **Probability** gives a measure of how likely an event is to happen.

PROBABILITY: AN OVERVIEW

When we toss a coin, there are two possible outcomes, heads or tails. We say that the probability of the coin landing heads is $\frac{1}{2}$ (the probability of the coin landing tails is the same). If we were to toss a coin 100 times, we would *expect* heads to come up $100 \times \frac{1}{2} = 50$ times. Note that we would not be surprised if the number was not exactly 50, although we would expect it to be round about 50.

We use the following definition for the probability of an event happening:

Probability of an event happening $= \dfrac{\text{Number of favourable outcomes}}{\text{Total number of outcomes}}$

EXAMPLE

A die is rolled. What is the probability that it shows a number greater than 4?

SOLUTION

The total number of outcomes is 6 (1, 2, 3, 4, 5 or 6).

The number of favourable outcomes is 2 (5 or 6).

Hence the probability $= \frac{2}{6}$ or $\frac{1}{3}$

If an event is *impossible*, for example scoring 7 when you roll a normal die, its probability is 0. If an event is *certain*, for example that tomorrow will be Sunday if today is Saturday, its probability is 1. The probability of any other event happening is somewhere between 0 and 1. Probabilities can *never* have either negative values or values greater than 1.

Probabilities are sometimes shown on a *probability line*.

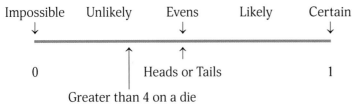

Of course, the probability of some events happening is extremely unlikely. The chance of buying the winning jackpot ticket in the lottery is 1 in 14 million. Other unlikely, but by no means impossible, events include damage to property caused by floods or gales, burglary to a property or being involved in a car accident. Because of the real risk of such events happening, people take out insurance to guard against them. Insurance companies employ experts to calculate the probability of such eventualities and charge customers premiums for insurance accordingly. It is a fact that young drivers looking to insure their first car will face very high insurance charges due to high road accident figures for young drivers.

COURSE IDEA

Investigate the statistics for accidents involving young drivers and find out things which young drivers could do in order to reduce insurance costs.

We will now look at some slightly more difficult examples on probability.

EXAMPLE

The table shows the results of a survey of fifth year pupils at a school.

	Wearing school uniform	Not wearing school uniform
Girls	30	20
Boys	28	22

What is the probability that a fifth year pupil, chosen at random from this sample, will be a girl wearing a school uniform?

Express your answer as a fraction in its simplest form.

SOLUTION

Start by finding the total number of pupils:
$30 + 20 + 28 + 22 = 100$

$$\text{Probability} = \frac{\text{Number of favourable outcomes}}{\text{Total number of outcomes}} = \frac{30}{100} = \frac{3}{10}$$

EXAMPLE

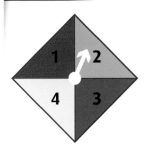

The arrow on a spinner is spun *twice* in an experiment and allowed to come to rest. The table below shows some of the possible outcomes.

2nd spin \ 1st spin	1	2	3	4
1	1, 1	2, 1		
2	1, 2			
3	1, 3			
4				

By copying and completing the table, find the probability that the **total** score for the two spins is 4.

SOLUTION

2nd spin \ 1st spin	1	2	3	4
1	1, 1	2, 1	3, 1	4, 1
2	1, 2	2, 2	3, 2	4, 2
3	1, 3	2, 3	3, 3	4, 3
4	1, 4	2, 4	3, 4	4, 4

Hence probability of a total score of $4 = \dfrac{3}{16}$

 DON'T FORGET

Although we have expressed probabilities as fractions in this section, and it is recommended that you do likewise, probabilities can appear in other forms. For example $\frac{1}{2}$ could be written as 0·5 or 50% or as a 1 in 2 chance. Do not write as 1:2 however.

THINGS TO DO AND THINK ABOUT

A letter is chosen at random from the word **MATHEMATICS**. What is the probability that it is M?

RELATIONSHIPS

THE STRAIGHT LINE

THE EQUATION OF A STRAIGHT LINE

When a straight line is drawn on a coordinate grid, then the straight line has an *equation*.

The equation connects the coordinates of all the points lying on the line. Consider the line shown in the diagram which passes through the points (0, 0), (1, 2), (2, 4), (3, 6), (4, 8) and so on. We can see that for every point on this line, the second coordinate (the y-coordinate) is double the first coordinate (the x-coordinate). Hence the equation of the straight line shown is $y = 2x$.

DRAWING A STRAIGHT LINE

You should be able to draw a straight line on a coordinate grid given its equation. To do this, you should replace selected values of x in the equation to find corresponding values of y. The straight line can then be drawn by plotting points and joining them.

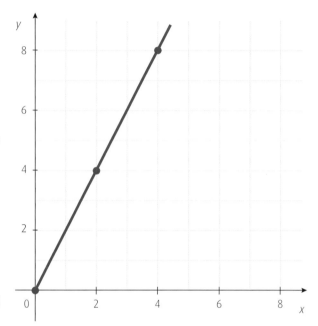

(a) Complete the table below for
$y = 0{\cdot}5x + 2$

x	0	2	4	6
y				

(b) Draw the line $y = 0{\cdot}5x + 2$

> BEWARE: If you complete a table of points, plot the points and then find that the points are not in a straight line, then there is an error somewhere in your working. In this case, check your working carefully until you identify the error. Do not join points if this leads to more than one line in your diagram.

(a) When $x = 0$, $y = 0{\cdot}5 \times 0 + 2 = 0 + 2 = 2$
When $x = 2$, $y = 0{\cdot}5 \times 2 + 2 = 1 + 2 = 3$
When $x = 4$, $y = 0{\cdot}5 \times 4 + 2 = 2 + 2 = 4$
When $x = 6$, $y = 0{\cdot}5 \times 6 + 2 = 3 + 2 = 5$

Therefore the points (0,2), (2,3), (4,4) and (6,5) all lie on the line.

x	0	2	4	6
y	2	3	4	5

(b)

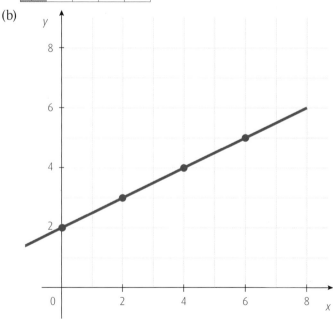

STRAIGHT LINES PARALLEL TO THE *x*- AND *y*-AXES

Write down the equation of lines (a) and (b) shown in the diagram below.

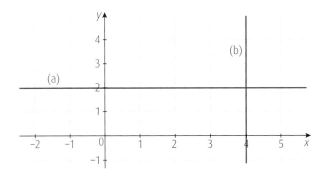

Line (a) passes through the points (–2,2), (–1,2), (0,2), (1,2), (2,2), (3,2), (4,2), (5, 2) and so on. In all of these points, the *y*-coordinate is 2, hence the equation of the line is $y = 2$.

In a similar way, we can show that the equation of line (b) is $x = 4$.

DON'T FORGET

Lines parallel to the axes have simple equations, for example $y = 2$ and $x = 3$. Be careful not to mix up *x* and *y*.

SUMMARY

- Equations such as $y = 2$, $x = 5$, $x = -3$, $y = 0$ and so on are all equations of straight lines.
- Horizontal lines are parallel to the *x*-axis and have an equation in the form 'y = a number'.
- Vertical lines are parallel to the *y*-axis and have an equation in the form 'x = a number'.

For example, the line parallel to the *y*-axis passing through the point (3,2) has equation $x = 3$.

A FORMULA FOR THE EQUATION OF A STRAIGHT LINE

The equation of a straight line is often given using the formula $y = mx + c$, where *m* represents the gradient of the straight line and *c* is the *y*-intercept. The *y*-intercept tells us where the straight line crosses the *y*-axis. In other words, if a straight line crosses the *y*-axis at the point $(0, c)$, then *c* is the *y*-intercept.

A straight line has equation $y = 3x + 4$.

(a) What is the gradient of the straight line?
(b) What are the coordinates of the point where it crosses the *y*-axis?

(a) The gradient is 3.
(b) It crosses the *y*-axis at the point (0,4).

Remind yourself how to find the gradient of a straight line. Check that the line $y = 0.5x + 2$ from the earlier example has gradient 0·5 and crosses the *y*-axis at the point (0,2). Investigate how the formula $y = mx + c$ can be used to check whether a straight line has been drawn correctly. This could be useful if the situation mentioned earlier in the 'BEWARE' note arises.

 THINGS TO DO AND THINK ABOUT

(a) Complete the table below for $y = 2x - 1$

x	1	2	3	4
y				

(b) Draw the line $y = 2x - 1$

EQUATIONS

WHAT IS AN EQUATION?

In the previous chapter we learned what an expression was in mathematics. An example of an expression is $3x + 5$. An example of an **equation** however could be $3x + 5 = 11$. An equation is a sentence with the verb 'is equal to' in it. The equation shown has one variable, namely x. If we are asked to solve the equation, then we have to find the value of x. The answer is called the solution of the equation and should be given in the form '$x =$'. Can you solve the above equation?

The equations you will meet in National 4 Mathematics are called linear equations. In a linear equation, the terms are either constants (numbers) or the product of a constant and a variable. Well done if you solved the linear equation given above. The solution is $x = 2$.

SOLVING EQUATIONS

You may have been able to solve the above equation using mental maths; however this will not always be possible. In fact, you will be expected to solve equations *algebraically*, and that means that you must always show your working indicating the processes required to arrive at the solution. To do this, we can think of an equation as a set of balanced scales. In order to keep a set of scales balanced, whatever we do to one side of the scales, we must do the same to the other side of the scales. Doing the same means either adding the same amount to both sides or subtracting the same amount from both sides or multiplying or dividing both sides by the same number.

EXAMPLE

Consider the equation $3x + 5 = 11$ mentioned above.

SOLUTION

We would set out working for this equation as follows:

$$
\begin{array}{rcll}
3x \ + \ 5 &=& 11 & \\
-5 & & -5 & \text{(subtract 5 from each side of the equation)} \\
\Rightarrow \quad 3x &=& 6 & \\
\div 3 & & \div 3 & \text{(divide each side of the equation by 3)} \\
\Rightarrow \quad x &=& 2 &
\end{array}
$$

MORE DIFFICULT EQUATIONS

Many equations will have constant terms *and* terms involving the variable on both sides of the equals sign. In order to solve such equations, you should aim to end up with the variable terms on the left side and the constant terms on the right side. Whatever you do to one side of the equation, do the same to the other to keep things balanced.

EXAMPLE

Solve algebraically
$5x - 3 = 3x + 9$

SOLUTION

We would set out working for this equation as follows:

$$
\begin{array}{rcll}
5x \ - \ 3 &=& 3x \ + \ 9 & \\
+3 & & +3 & \text{(add 3 to each side of the equation)} \\
\Rightarrow \quad 5x &=& 3x \ + \ 12 & \\
-3x & & -3x & \text{(subtract } 3x \text{ from each side of the equation)} \\
\Rightarrow \quad 2x &=& 12 & \\
\div 2 & & \div 2 & \text{(divide each side of the equation by 2)} \\
\Rightarrow \quad x &=& 6 &
\end{array}
$$

JUST A WEE NOTE

The symbol $\Rightarrow$ can be used when one statement follows on logically from the previous statement, for example, $2x = 12 \Rightarrow x = 6$ means that if $2x = 12$ is true, it follows that $x = 6$. It is good to use this symbol but not essential.

Some examples may involve negative numbers. If unsure about this, check the section on Negative numbers on page 74.

EXAMPLE

Solve algebraically
$5x - 2 = -12$

SOLUTION

$$
\begin{array}{rccc}
& 5x & - \quad 2 & = & -12 \\
& & +2 & & +2 \quad \text{(add 2 to each side of the equation)} \\
\Rightarrow & & 5x & = & -10 \\
& & \div 5 & & \div 5 \quad \text{(divide each side of the equation by 5)} \\
\Rightarrow & & x & = & -2
\end{array}
$$

INEQUATIONS

An **inequation** is a sentence containing 'greater than' or 'less than' in it.
You must be aware of the following symbols.

- $x > 3$ means x is greater than 3
- $x < 3$ means x is less than 3
- $x \geqslant 3$ means x is greater than or equal to 3
- $x \leqslant 3$ means x is less than or equal to 3

The inequations in National 4 Mathematics can be solved in the same way as equations.

EXAMPLE

Solve the inequation $3x + 5 > 11$

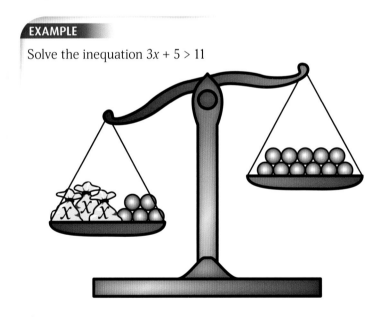

SOLUTION

$$
\begin{array}{rccc}
& 3x \quad + \quad 5 & > & 11 \\
& -5 & & -5 \\
\Rightarrow & 3x & > & 6 \\
& \div 3 & & \div 3 \\
\Rightarrow & x & > & 2
\end{array}
$$

THINGS TO DO AND THINK ABOUT

1. Solve algebraically
   (a) $2x + 7 = 11$

   (b) $5x - 2 = 4x + 6$

   (c) $8x + 4 = 2x + 40$

   (d) $5x + 3 = -7$

2. Solve the inequation $3y + 4 \leqslant 19$

CHANGING THE SUBJECT OF A FORMULA

WHAT IS THE SUBJECT OF A FORMULA?

We have already met numerous formulae in earlier sections. Well known examples include $A = lb$, $C = \pi d$ and $V = lbh$. If you consider these three examples, the subjects are A, C and V respectively. It is very useful sometimes to *change* the subject of a formula. For example, suppose we know the circumference of a circle and want to calculate the diameter, then it would help if we could change the subject of the formula $C = \pi d$ from C to d. We shall investigate how to do this.

CHANGING THE SUBJECT

We shall use the ideas from the previous section on equations to change the subject of a formula. Formulae are basically equations involving symbols and variables, so the ideas on balancing can be applied. We shall change the subject of a formula by showing the solution alongside a similar linear equation of the type we have already seen.

EXAMPLE

(a) Solve the equation $3x = 12$
(b) Change the subject of the formula $C = \pi d$ to d

SOLUTION

(a)
$$3x = 12$$
$$\div 3 \qquad \div 3$$
$$\Rightarrow x = 4$$

(b)
$$C = \pi d$$
$$\Rightarrow \pi d = C \quad \text{(swap the sides around)}$$
$$\div \pi \qquad \div \pi \quad \text{(divide each side by } \pi)$$
$$\Rightarrow d = \frac{C}{\pi}$$

EXAMPLE

The total salary, s pounds, of an employee earning a basic wage, b pounds, and commission, c pounds, is given by the formula

$s = b + c$.

Change the subject of the formula to b

SOLUTION

$$s = b + c$$
$$\Rightarrow b + c = s \quad \text{(swap the sides around)}$$
$$-c \qquad -c \quad \text{(subtract } c \text{ from each side)}$$
$$\Rightarrow b = s - c$$

EXAMPLE

Change the subject of the formula $V = lbh$ to b

SOLUTION

$$V = lbh$$
$$\Rightarrow lbh = V \quad \text{(swap the sides around)}$$
$$\div lh \qquad \div lh \quad \text{(divide each side by } lh)$$
$$\Rightarrow b = \frac{V}{lh}$$

BEWARE: A common wrong answer to the above question is $b = V - lh$. However as there is no addition (+ sign) in the original formula, there cannot be a subtraction (- sign) when you change the subject of the formula.

EXAMPLE

Change the subject of the formula $P = 5m - 3$ to m. (This is an example of a two-step formula)

SOLUTION

$$P = 5m - 3$$
$$\Rightarrow 5m - 3 = P \qquad \text{(swap the sides around)}$$
$$ +3 \qquad +3 \qquad \text{(add 3 to each side)}$$
$$\Rightarrow 5m = P + 3$$
$$ \div5 \qquad \div5 \qquad \text{(divide each side by 5)}$$
$$\Rightarrow m = \frac{P + 3}{5}$$

Now we shall return to an example from the section on Formulae on page 11 and find the solution this time by changing the subject of the formula.

EXAMPLE

The formula for converting degrees Celsius to degrees Fahrenheit is

$$F = 1{\cdot}8C + 32$$

By changing the subject of the formula to C, convert 50° Fahrenheit to degrees Celsius.

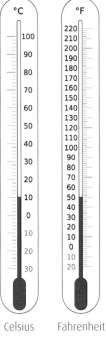

Celsius   Fahrenheit

SOLUTION

$$F = 1{\cdot}8C + 32$$
$$\Rightarrow 1{\cdot}8C + 32 = F \qquad \text{(swap sides around)}$$
$$\phantom{\Rightarrow 1{\cdot}8C} -32 \qquad -32 \qquad \text{(subtract 32 from each side)}$$
$$\Rightarrow \phantom{1{\cdot}8C} 1{\cdot}8C = F - 32$$
$$\phantom{\Rightarrow 1{\cdot}8C} \div1{\cdot}8 \qquad \div1{\cdot}8 \qquad \text{(divide each side by 1{\cdot}8)}$$
$$\Rightarrow \phantom{1{\cdot}8C} C = \frac{F - 32}{1{\cdot}8}$$

Hence $C = \dfrac{F - 32}{1{\cdot}8} = \dfrac{50 - 32}{1{\cdot}8} = \dfrac{18}{1{\cdot}8} = 10$.

So the solution is 10° Celsius.

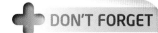 DON'T FORGET

Remember that the method used to change the subject of a formula is the same as the method used to solve an equation.

COURSE IDEA

Try to change the subject of the formula $A = \pi r^2$ to r. This is more difficult. Then use your solution to find the radius of a circle whose area is 616 square centimetres. The correct solution is 14 centimetres. Well done if you got this correct.

THINGS TO DO AND THINK ABOUT

1. Change the subject of the formula $N = G - D$ to G.

2. Change the subject of the formula $V = IR$ to I.

3. Change the subject of the formula $M = \frac{Y}{12}$ to Y.

4. Change the subject of the formula $s = 4c + 1$ to c.

5. Change the subject of the formula $A = Py - Q$ to y.

THE THEOREM OF PYTHAGORAS

PYTHAGORAS THEOREM: AN OVERVIEW

Pythagoras was a Greek mathematician who lived in the 6th century BC. He discovered an important fact about right-angled triangles. It is called the **Theorem of Pythagoras**. In words, the theorem states that:

In a right-angled triangle, the square on the hypotenuse is equal to the sum of the squares on the other two sides.

The **hypotenuse** is the longest side in a right-angled triangle and is opposite the right angle. A theorem is a true statement which can be expressed in words or by a formula. The formula for the Theorem of Pythagoras is shown below.

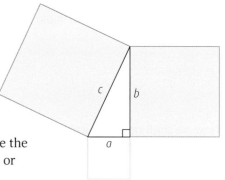

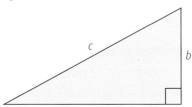

$$a^2 + b^2 = c^2$$

JUST A WEE NOTE

In most examples the theorem is used in the form $c^2 = a^2 + b^2$.

EXAMPLE

Find x.

SOLUTION

Using the Theorem of Pythagoras,
$$x^2 = 12^2 + 5^2 = 144 + 25 = 169$$
Hence $x = \sqrt{169} = 13\,\text{cm}$

JUST A WEE NOTE

Note that as x is the hypotenuse, we square and add the other two sides and then complete the calculation by finding the square root of the total.

EXAMPLE

Triangle PQR is right-angled with measurements shown below.

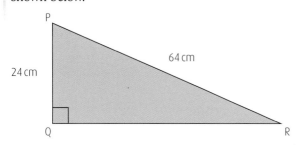

Calculate the length of QR. Round your answer to the nearest centimetre.

SOLUTION

Using the Theorem of Pythagoras,

$PR^2 = QR^2 + PQ^2$, hence $QR^2 = PR^2 - PQ^2$

$QR^2 = 64^2 - 24^2 = 4096 - 576 = 3520$

Hence $QR = \sqrt{3520} = 59\,\text{cm}$ (to the nearest centimetre)

JUST A WEE NOTE

Note that as QR is one of the shorter sides, we square and subtract the other two sides and then complete the calculation by finding the square root of the answer.

DON'T FORGET

If you are using the Theorem of Pythagoras to find the length of the hypotenuse, square and add the other two sides. If you are using the Theorem to find the length of one of the two shorter sides, square and subtract the other two sides.

COORDINATES

We shall refer back to an earlier example on coordinates in the section on Gradient (on page 15) to show how the Theorem of Pythagoras can be applied to examples involving coordinates.

EXAMPLE

Find the length of the straight line joining the points A (–4, 5) and B (3, 2).

SOLUTION

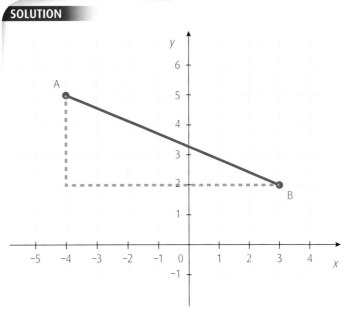

Using the Theorem of Pythagoras, $AB^2 = 7^2 + 3^2 = 49 + 9 = 58$

Hence $AB = \sqrt{58} = 7{\cdot}6$ (correct to one decimal place)

DON'T FORGET

The Theorem of Pythagoras is an important topic in mathematics and is very likely to be tested in assessments, so make sure you are confident about using it.

COURSE IDEA

Sometimes the three sides in a right-angled triangle are whole numbers, for example 5, 12 and 13 in the first example. Such groups of numbers are called *Pythagorean triples*. It is very useful to be aware of the most common triples. By completing the table below, you will discover some of them. The table is completed by taking two starting numbers, a and b such that $a > b$, for example $(a = 2, b = 1)$, $(a = 3, b = 1)$, $(a = 3, b = 2)$, $(a = 4, b = 1)$, $(a = 4, b = 2)$, etc.

a	b	$a^2 + b^2$	$a^2 - b^2$	$2ab$	*Pythagorean triple*
2	1	$2^2 + 1^2 = 4 + 1 = 5$	$2^2 - 1^2 = 4 - 1 = 3$	$2 \times 2 \times 1 = 4$	**3, 4, 5**
3	1	$3^2 + 1^2 = 9 + 1 = 10$	$3^2 - 1^2 = 9 - 1 = 8$	$2 \times 3 \times 1 = 6$	**6, 8, 10**
3	2				
4	1				
4	2				
4	3				

THINGS TO DO AND THINK ABOUT

A component in a piece of machinery is in the shape of a right-angled triangle.

Calculate the length of the side marked *x*.

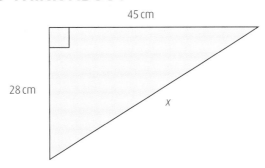

45 cm

28 cm

x

SCALE FACTORS

WHAT IS A SCALE FACTOR?

When a shape is enlarged (or reduced), the size of the enlargement (or reduction) is described by its **scale factor**. For example, a scale factor of 2 means that all the lengths in the new shape are twice the lengths of those in the original shape, whereas a scale factor of $\frac{1}{2}$ means that all the lengths in the new shape are half the lengths of those in the original shape. When one shape is an enlargement or reduction of another shape, the two shapes are said to be **similar**. In mathematics, we say that two shapes are similar if both shapes are equiangular, that is they have equal angles *and* their corresponding sides have been enlarged or reduced by the same scale factor.

EXAMPLE

Rectangle A has been enlarged in the diagram below. What is the scale factor of the enlargement?

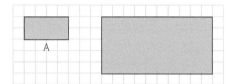

SOLUTION

Compare lengths $\left(\frac{10}{4}\right)$ or breadths $\left(\frac{5}{2}\right)$. The scale factor would be given in its simplest form, that is $\frac{5}{2}$ or in decimal form, 2·5.

DON'T FORGET

When finding the scale factor for an enlargement, the larger number goes on the numerator of the fraction. When finding the scale factor for a reduction, the smaller number goes on the numerator of the fraction.

ENLARGING AND REDUCING

You will be expected to enlarge or reduce more complex shapes, involving sloping lines, using fractional scale factors. It is recommended that you deal with any horizontal or vertical lines first.

EXAMPLE

Draw an enlargement of the given shape using a scale factor of $\frac{3}{2}$

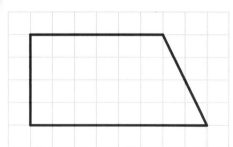

SOLUTION

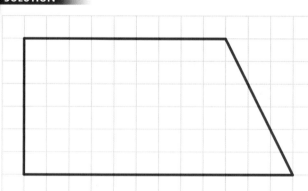

METHOD

Find the lengths of the vertical line (4) and the two horizontal lines (6) and (8) in the original diagram by counting. Then multiply each by $\frac{3}{2}$ as follows:

$$4 \times \tfrac{3}{2} = 4 \times 3 \div 2 = 6, \quad 6 \times \tfrac{3}{2} = 6 \times 3 \div 2 = 9, \quad 8 \times \tfrac{3}{2} = 8 \times 3 \div 2 = 12$$

The vertical line and the two horizontal lines are drawn first in the enlargement. This means that the more difficult sloping line can easily be joined in.

DON'T FORGET

Always draw horizontal and vertical lines first when enlarging or reducing a shape. If there are no horizontal and vertical lines, it is a good idea to draw a rectangle around the entire shape and enlarge or reduce the rectangle to get started.

SCALE DRAWINGS

Scale factors are used in maps and plans. You might see a scale on a plan of a kitchen written as 1:100. This means that the *actual* lengths in the kitchen are 100 times bigger than those on the plan. In other words, 1 centimetre on the plan would represent an actual distance of 100 centimetres or 1 metre. This scale could also be written as '1 cm represents 1 m'. You may be asked to make a scale drawing.

EXAMPLE

A ship has to visit some oil rigs in the North Sea. The ship departs from its base. It sails 40 kilometres due east, then 20 kilometres due south and then 30 kilometres due west to oil rig Omega.

  (a) Make a scale drawing to show the ship's journey.
      Use a scale of 1 centimetre represents 5 kilometres.

  (b) The ship then returns directly from oil rig Omega to its base.
      Use your scale drawing to find the distance of this return journey.

SOLUTION

Distances of 40 ÷ 5 = 8, 20 ÷ 5 = 4, 30 ÷ 5 = 6 should be used (in centimetres).

(a)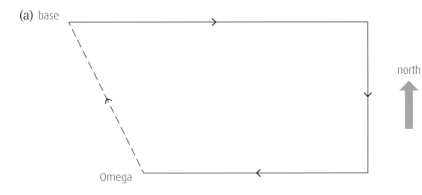

Use a ruler to measure the distance from Omega to base (approximately 4·5 centimetres) then multiply by 5 for the solution in kilometres, that is 5 × 4·5 = 22·5 km approximately.

For practice, try to calculate the distance from Omega to base using the Theorem of Pythagoras. Your solution should be close to 22·5 metres.

COURSE IDEA

Consider a rectangle 4 centimetres by 3 centimetres which is enlarged using a scale factor of 2. Do you think that the area of the enlargement will be two times bigger? Well, the original rectangle has an area of 4 × 3 = 12 cm² and the enlargement will have an area of 8 × 6 = 48 cm². Instead of being two times bigger, you can see that the area is 48 ÷ 12 = 4 times bigger. Think of enlarging the original rectangle using a scale factor of 3 and go on to investigate the relationship between similar shapes and their areas.

THINGS TO DO AND THINK ABOUT

Draw an enlargement of the given octagon using a scale factor of $\frac{3}{2}$.

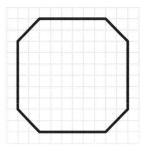

ANGLES

MEASURING ANGLES

You should be able to use a protractor to measure angles. Study the example below.

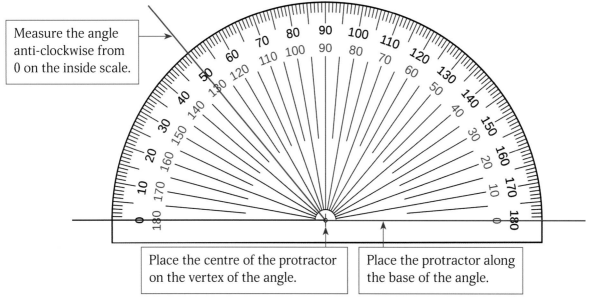

Measure the angle anti-clockwise from 0 on the inside scale.

Place the centre of the protractor on the vertex of the angle.

Place the protractor along the base of the angle.

The **vertex** of an angle is the corner where the two arms of the angle meet.
The angle shown in the above diagram is 130°.

TYPES OF ANGLES

You should already know that an **acute** angle is less than 90°, a right angle = 90°, an **obtuse** angle is greater than 90° and less than 180°, a straight angle = 180°, a **reflex** angle is greater than 180° and less than 360° and a complete turn = 360°.

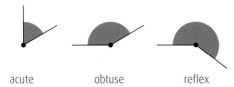

acute   obtuse   reflex

Other important types of angles are mentioned below:

(i) **Vertically opposite angles** are equal. They are formed when two straight lines intersect. They get their name from the fact that they share the same vertex.

(ii) **Corresponding angles** are equal. They also occur when a straight line intersects parallel lines. They form an F-shape.

(iii) **Alternate angles** are equal. They occur when a straight line intersects parallel lines. They form a Z-shape.

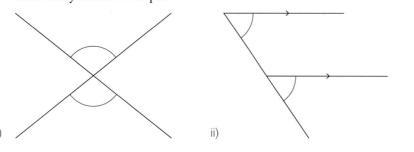

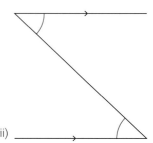

i)   ii)   iii)

ANGLE PROBLEMS

We shall now look at some angle problems. Remember that the three angles in a triangle always add up to 180°.

EXAMPLE

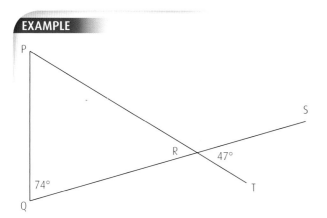

Calculate the size of angle QPR.

SOLUTION

$\angle$PRQ = 47° (vertically opposite angles)

$\angle$QPR = 180° − (74 + 47)° = 59° (sum of angles in a triangle = 180°)

EXAMPLE

STUV is a parallelogram. Side UV has been extended to point W.

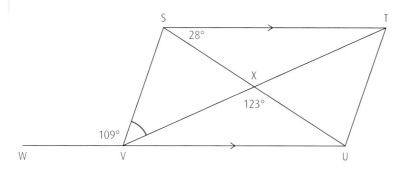

Angle SVW = 109°, angle UXV = 123° and angle TSX = 28°.

Calculate the size of angle SVX.

SOLUTION

We can use alternate angles (Z-shape) as ST is parallel to VU (indicated by arrows).

Hence $\angle$XUV = 28°
(alternate angles: Z-shape)

$\angle$XVU = 180° − (123 + 28)° = 29°
(sum of angles in triangle = 180°)

$\angle$SVX = 180° − (109 + 29)° = 42°
($\angle$UVW = 180° as it is a straight angle)

THINGS TO DO AND THINK ABOUT

ABCD is a trapezium with AB parallel to DC.

Angle ABD = 36° and angle BCD = 60°.

Calculate the size of angle CBD.

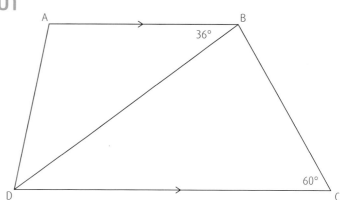

MORE ANGLES

THE CIRCLE

You should know that an **isosceles triangle** is a triangle with two equal sides. The two angles opposite the equal sides are also equal.

Isosceles triangles often occur in circles when a triangle is drawn with two of its sides being radii (plural of radius). Look out for isosceles triangles in circles as they are often the key to solving angle problems.

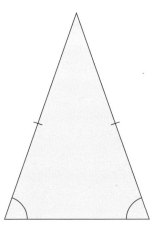

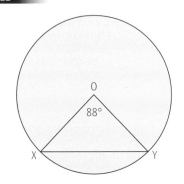

X and Y are points on the circumference of the circle, centre O. Angle XOY = 88°.

Calculate the size of angle OYX.

SOLUTION

∠OYX + ∠OXY = (180 − 88)° = 92°
(sum of angles in a triangle = 180°)

∠OYX = ∠OXY as triangle OXY is isosceles because OX = OY (both are radii)

Hence ∠OXY = (92 ÷ 2)° = 46°

COURSE IDEA

Investigate the properties of an equilateral triangle. Find out about its sides and angles and look into its properties regarding both line and rotational symmetry.

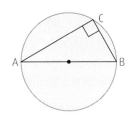

THE ANGLE IN A SEMI-CIRCLE

If AB is a diameter of a circle and C is a point on the circumference, then angle ACB is a right angle (90°). We usually say that 'the angle in a semi-circle is a right angle'.

EXAMPLE

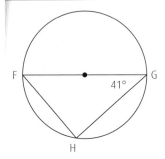

FG is the diameter of the circle above. Angle FGH = 41°.

Calculate the size of angle GFH.

SOLUTION

∠FHG = 90°
(angle in a semi-circle)

∠GFH = 180° − (90 + 41)° = 49°
(sum of angles in a triangle = 180°)

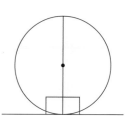

TANGENTS

A **tangent** to a circle is a straight line which touches the circle at one point only. This point is called the point of contact. At the point of contact, the tangent is **perpendicular** (at right angles) to the radius (or diameter).

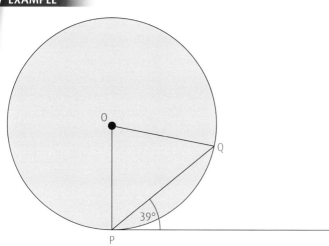

PT is a tangent at point P to the circle, centre O.
Angle TPQ = 39°.

Calculate the size of angle POQ.

∠OPT = 90°   (angle between tangent and radius).

∠OPQ = 90° − 39° = 51°

∠OQP = 51° as triangle OPQ is isosceles because OP = OQ (both are radii)

∠POQ = 180° − (51 + 51)° = 78°   (sum of angles in triangle = 180°)

DON'T FORGET

Remember to fill in the sizes of any angles you calculate in this type of problem on the diagram.

THINGS TO DO AND THINK ABOUT

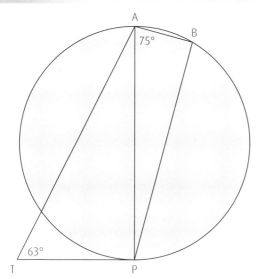

In the above diagram, AP is a diameter of the circle.

TP is a tangent to the circle at P.

Point B lies on the circumference of the circle.

Angle ATP = 63° and angle PAB = 75°.

Calculate the size of angle BAT.

TRIGONOMETRY

In the next two sections, we shall use trigonometry to calculate the length of sides and the sizes of angles in right-angled triangles. You will require a scientific calculator in order to use the important keys for trigonometry, namely sin, cos and tan (short for sine, cosine and tangent). These are referred to as trigonometric ratios as each is the ratio of two sides in a right-angled triangle.

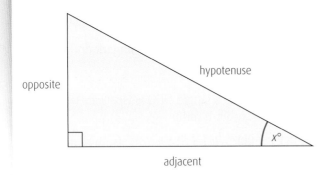

THE THREE SIDES

We have special names for the three sides in a right-angled triangle. We already know that the side opposite the right angle (the longest side in the triangle) is the hypotenuse. The side opposite the marked angle ($x°$ in the diagram) is called the opposite. The remaining side is called the adjacent where adjacent means next to the angle.

THE TRIGONOMETRIC RATIOS

Problems in triangles can be solved using the three ratios – sine, cosine and tangent. In assessments, the definitions of these ratios are given in a formula list, shown below.

$$\tan x° = \frac{\text{opposite}}{\text{adjacent}}$$

$$\sin x° = \frac{\text{opposite}}{\text{hypotenuse}}$$

$$\cos x° = \frac{\text{adjacent}}{\text{hypotenuse}}$$

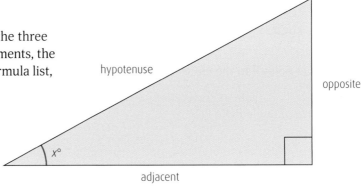

COURSE IDEA

Although you should always refer to the formulae list when doing assessments, there may be occasions when it is not available. A useful mnemonic to the formulae is the made up word SOHCAHTOA. Find out about how to use it as it will be helpful now (when you do not have the formula list available) and if you study trigonometry further in the future.

USING A CALCULATOR

EXAMPLE

(a) Find the value of cos 60°

(b) Given that sin $x° = 0.375$, find x to the nearest whole number.

SOLUTION

(a) 0·5   (Press cos 60 =).

(b) $x = \sin^{-1} 0.375 = 22.02431284 = 22$
(to the nearest whole number)
(Press SHIFT sin 0.375 =)

DON'T FORGET

Make sure you can do both (a) and (b). If you get incorrect solutions, it may be because your calculator is in the wrong mode, that is RAD or GRAD instead of DEG or D for degrees. Note too that some calculators use INV or 2nd F instead of SHIFT so you must understand your own calculator.

CALCULATING THE LENGTH OF A SIDE

We shall now investigate how to calculate the length of a side in a right-angled triangle using trigonometry, when you are given the size of an angle and the length of another side.

EXAMPLE

A telegraph pole is supported by a wire of length 12 metres. The wire makes an angle of 78° with the ground.

Calculate the height, h metres, of the telegraph pole.

DON'T FORGET

Always ensure that your calculator is in degree mode when you are doing a trigonometry problem.

SOLUTION

You should start by naming the three sides in the triangle.

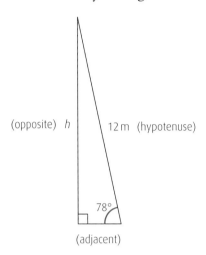

The given side is the *hypotenuse* and the side we wish to calculate is the *opposite* as it is opposite the marked angle, 78°. Check the formula list for hypotenuse and opposite and you will see that we must use sine as sine $= \dfrac{\text{opposite}}{\text{hypotenuse}}$

$\sin 78° = \dfrac{h}{12}$ (now multiply both sides by 12).

$\Rightarrow h = 12 \times \sin 78° = 11{\cdot}73777121$

Hence the height of the telegraph pole is 11·74 m (correct to two decimal places)

THINGS TO DO AND THINK ABOUT

1. Find the length of the side marked x in the following triangles. Give your answers correct to one decimal place.

   (a)

   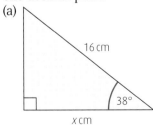

   (b)

   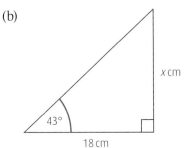

2. Rashid is going to measure the height of his school building. At a distance of 45 metres from the foot of the building he measures the angle of elevation as 50° to the top of the building as shown in the diagram.

   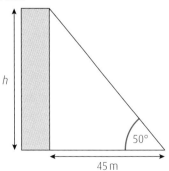

   Calculate the height, h metres, of the building to the nearest metre.

MORE TRIGONOMETRY

CALCULATING THE SIZE OF AN ANGLE

We shall now investigate how to calculate the size of an angle in a right-angled triangle using trigonometry when you are given the lengths of two sides.

EXAMPLE

Calculate the size of the angle marked $x°$ in the diagram below.
Give your answer to the nearest degree.

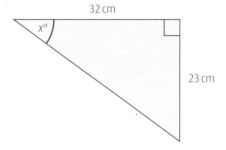

SOLUTION

Again, you should start by naming the three sides in the triangle.

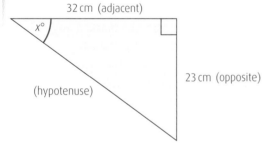

The given sides are the *opposite* and the *adjacent*. Check the formula list for opposite and adjacent and you will see that we must use tangent.

$$\tan x° = \frac{\text{opposite}}{\text{adjacent}} = \frac{23}{32}$$

Now press SHIFT tan (23 ÷ 32) =

This will lead to $x = \tan^{-1}(23 ÷ 32) = 35{\cdot}7066914$

Hence $x = 36$ (to the nearest degree)

DON'T FORGET

As with the Theorem of Pythagoras, it is very likely that trigonometry will be tested in assessments, so make sure you are confident about the examples shown. Look out for questions with diagrams involving triangles asking you to calculate lengths and/or angles. Don't confuse questions on the Theorem of Pythagoras with trigonometry. In trigonometry diagrams, there will be angles, for example 78° or $x°$.

HARDER EXAMPLES

EXAMPLE

A ladder, 5 metres long, is placed against a wall by a window cleaner. The ladder makes an angle of $x°$ with the ground.

The base of the ladder has been placed 1·2 metres from the foot of the wall as shown in the diagram.

For safety reasons, angle $x°$ should be greater than 75° and less than 78°.

Is it safe for the window cleaner to use the ladder in this position? Justify your answer.

SOLUTION

Check that sides 5 m and 1·2 m are the hypotenuse and adjacent respectively, so we must use cosine.

$$\cos x^o = \frac{\text{adjacent}}{\text{hypotenuse}} = \frac{1 \cdot 2}{5}$$

Now press SHIFT cos (1·2 ÷ 5) =, leading to $\cos^{-1}$ (1·2 ÷ 5) = 76·11345964

It is safe to use the ladder because 76·11345964° is greater than 75° and less than 78°.

DON'T FORGET

If you are asked to justify your answer, this means that you should answer in words and refer to the numerical values in your calculations and in any conditions stated in the question.

Finally, we shall consider a trickier example on finding the length of a side because the side we wish to calculate is on the *denominator* of the fraction in the trigonometric ratio.

EXAMPLE

Find the length of the side marked x in the triangle shown below.

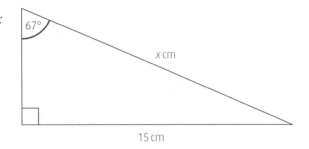

SOLUTION

Check that sides x cm and 15 cm are the hypotenuse and opposite respectively (be careful as 15 cm is opposite the marked angle, 67°), so we must use sine.

$$\sin 67^o = \frac{\text{opposite}}{\text{hypotenuse}} = \frac{15}{x}$$

$\Rightarrow \sin 67^o \times x = 15$ (multiply both sides by x)

$\Rightarrow x = 15/\sin 67^o$ (divide both sides by $\sin 67^o$)

$\Rightarrow x = 16 \cdot 29540566$ Hence $x = 16 \cdot 3$ cm (correct to one decimal place)

COURSE IDEA

Investigate the connection between gradient (=Vertical height /Horizontal distance) and the tangent ratio (=opposite /adjacent) and consider how health and safety issues, such as the use of ladders at work, could involve the use of trigonometry.

 THINGS TO DO AND THINK ABOUT

Find the size of the angle marked x^o in the following triangle.

Give your answer correct to one decimal place.

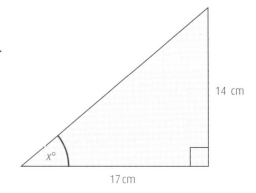

SCATTERGRAPHS

DRAWING A SCATTERGRAPH

A **scattergraph** is a statistical diagram which can be used to compare two sets of data. The scattergraph is a set of points plotted on a coordinate grid. We can use it to look for connections between the two data sets on the graph.

EXAMPLE

A school is carrying out a survey to find the relationship between student performance in music and mathematics. The percentage marks of a group of S1 students in their music and mathematics exams are listed below.

Student	A	B	C	D	E	F	G	H	I	J
Music mark	52	35	63	47	72	32	28	57	69	75
Mathematics mark	61	32	58	42	77	59	23	80	59	81

Draw a scattergraph to illustrate these marks.

SOLUTION

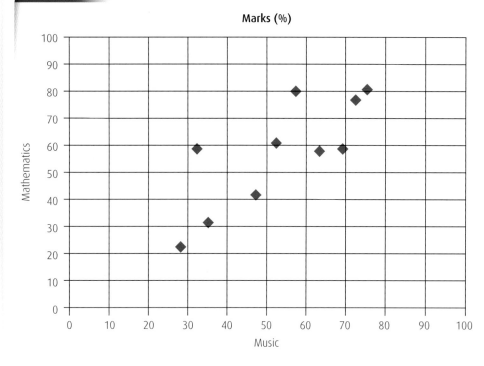

ADVICE

Always use clear scales on both axes. Label both axes and insert a heading above the scattergraph. The *first* set of data in the table (music marks in this case) should be plotted on the horizontal axis. Note that in assessments, you may be provided with a ready-made grid on which you can plot the points. When plotting the points, (52, 61), (35, 32) and so on, take care as you may have to estimate the position of some in-between values.

THE BEST-FITTING LINE

When a scattergraph has been drawn, it is sometimes possible to draw a best-fitting line which follows the direction of the points. It can also be called the line of best fit. This line can be used to estimate other data.

EXAMPLE

(a) Draw a best-fitting line on the above scattergraph.

(b) Student K scored 50 marks for music. Use the best-fitting line to estimate the student's mark for mathematics.

(a)

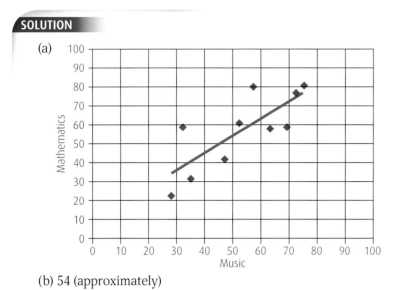

(b) 54 (approximately)

CORRELATION

A scattergraph helps us to see the connection or **correlation** between two data sets.

If, as one data set increases, the other also increases, we say there is a positive correlation, for example, height and weight.

If, as one data set increases, the other decreases, we say there is a negative correlation, for example, the number of flu jabs given and the number of people catching flu.

If a scattergraph shows a cloud pattern, there is no correlation, for example, height and salary.

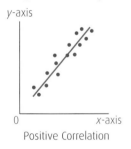

Positive Correlation

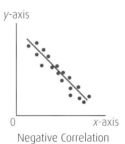

Negative Correlation

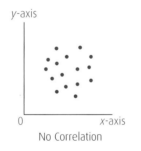

No Correlation

What type of correlation is shown in the scattergraph for music and mathematics?

Positive correlation (note that the gradient of the best-fitting line is positive).

COURSE IDEA

Find out how to design a scattergraph on a computer, for example, using Microsoft Excel.

THINGS TO DO AND THINK ABOUT

Refer to the earlier scattergraph and the best-fitting line shown above.

1. Student L scored 30 marks for music. Estimate the student's mark for mathematics.
2. Student M scored 70 marks for mathematics. Estimate the student's mark for music.

ADVICE

When drawing a best-fitting line, the line should follow the general path of the points. Draw the line so that it has the same number of points above as it has below, as far as possible. For example, the best-fitting line here has five points above and five points below. In order to achieve this it is a good idea to use a transparent ruler to draw the line. Note that the line may pass through some of the points. The line can also be extended if necessary. It does not need to go through the origin. There are different acceptable answers to this question.

 **DON'T FORGET**

Use a transparent ruler to draw a best-fitting line as it is easy to check that the number of points above and below the line is roughly the same.

NUMERACY

WHOLE NUMBERS

WORKING WITHOUT A CALCULATOR

After studying this section, without a calculator you should be able to

- add and subtract whole numbers
- multiply whole numbers by a single digit whole number or by 10 or 100 or 1000
- divide whole numbers of any size by a single digit whole number or by 10 or 100 or 1000

It is important that you can carry out the four basic operations (add, subtract, multiply and divide) without a calculator. Cover up the solutions to the four examples below and concentrate on getting all four correct.

EXAMPLE

(a) The number of pupils in each year of a secondary school is given below.
172   189   181   136   82   38
Find the total number of pupils in the school.

(b) David has a collection of 605 football programmes. He sells 187 of the programmes. How many remain in his collection?

(c) A political party is going to print copies of a document containing nine pages. If 685 copies of the document are printed, how many pages will be printed altogether?

(d) A supermarket chain has received a delivery of 623 grapefruits to its depot. An equal number of grapefruits are to be sent to each of its seven local branches. How many grapefruits will each branch receive?

SOLUTION

  (a) 798   (b) 418   (c) 6165   (d) 89

DON'T FORGET

While these examples were not particularly difficult, it is not uncommon for students to make careless or unforced errors. Due to the nature of your assessments in National 4 Mathematics, careless errors could mean that you do not pass an assessment. Therefore treat all examples with respect and check your solutions carefully.

WORKING WITH A CALCULATOR

You may use a calculator for the following examples, which are more difficult. Read each question, carefully, decide on a strategy and remember to show your working clearly.

EXAMPLE

A cardboard box is used to transport bottles of wine. It weighs 1275 grams when empty. It has to be filled with bottles of wine each weighing 880 grams for transportation. After the bottles of wine are put into the cardboard box, it weighs 14 475 grams.

How many wine bottles were put into the cardboard box?

SOLUTION

Weight of wine bottles = (14 475 − 1275) g = 13 200 g

Number of wine bottles = 13 200 ÷ 880 = 15

EXAMPLE

Darius is going to buy some equipment for his new kitchen. He sees the following offers in a department store.

KITCHEN EQUIPMENT
SPECIAL OFFERS

| Microwave | £45 | Blender | £40 |
| Kettle | £25 | Cafetière | £35 | Toaster | £20 |

Darius wants to buy **three** of these items. He can afford to spend £100.

One combination of items Darius can buy is shown in the table below.

Microwave £45	Blender £40	Kettle £25	Cafetière £35	Toaster £20	Total cost
✔		✔		✔	£90

Complete the table to show all the possible combinations that Darius can buy under the special offer.

SOLUTION

Microwave £45	Blender £40	Kettle £25	Cafetière £35	Toaster £20	Total cost
✔		✔		✔	£90
✔			✔	✔	£100
	✔	✔	✔		£100
	✔	✔		✔	£85
	✔		✔	✔	£95
		✔	✔	✔	£80

ADVICE

The example above involves extracting information from a table and adding money (whole numbers of pounds). Read the instructions carefully. Take a logical approach, that is start with £45 first, then move along the row checking possibilities so that none are missed out. Check all additions. Ensure that all five blank rows are filled in but watch out that you do not repeat a row, either the given one or one of your own entries. Questions of this type are straightforward but take time, so be patient.

COURSE IDEA

Practise mental mathematics involving whole numbers, for example you could try some multiplications such as 6 × 14 by doing 6 × (10 + 4) using the distributive law (see the section on Multiplying out brackets and Factorisation). Try similar examples mentally then check your answer on a calculator. As you practise you will improve and gain in confidence and you can move on to the other operations.

THINGS TO DO AND THINK ABOUT

A patient is prescribed medicine at her surgery. She has to take **three** 5 millilitre spoonfuls **twice** a day. She has to continue this dose for 12 days. The bottle contains 350 millilitres of medicine. Does the bottle contain enough medicine?

Justify your answer by calculation.

ROUNDING

APPROXIMATION

In real-life situations, numbers are often rounded to approximate numbers depending on the accuracy required. For example, the attendance at a football match is often rounded to the nearest thousand. If 73 802 spectators attended a match, we might be asked to round this to the nearest thousand.

| 73 000 | 73 500 | 73 802 | 74 000 |

We can see from the diagram that 73 802 ≈ 74 000 (to the nearest thousand).

If a number is halfway, we round up. For example 73 500 rounded to the nearest thousand is 74 000.

DECIMAL PLACES

When we need to measure a quantity very accurately, we can round to a given number of decimal places. Using decimal places (the number of figures that appear after the decimal point) is a common way of rounding.

For example 6·7538 has four decimal places. Suppose you were asked to round it to two decimal places.

- 6·7538 lies between 6·75 and 6·76 (each of which has two decimal places)
- look at the figure in the third decimal place (6·75**3**), that is 3
- since this figure is **less than 5**, do not round up, leave as 6·75

Check that 6·7538 rounded to one decimal place would be 6·8.

DON'T FORGET

If the next figure is 5 or more, round up. If the next figure is 4 or less, do not round up.

EXAMPLE

Round

(a) 6·372 to one decimal place

(b) 12·782345 to two decimal places.

SOLUTION

(a) 6·4

(b) 12·78

SIGNIFICANT FIGURES

Another method of rounding to an approximate number is to use significant figures (sig. figs for short). In a number, all figures are significant *except* zeros which are used simply to indicate the position of the decimal point. However, zeros in between other significant figures are themselves significant.

Hence, 589 346 has 6 significant figures; 715·28 has 5 significant figures; 4045 has 4 significant figures. Note that 0·00348 has only 3 significant figures as the zeros at the start indicate the position of the decimal point, whereas 0·003048 has 4 significant figures as the zero between the 3 and 4 is significant. A measurement such as 25.0 centimetres has 3 significant figures as the final zero tells you that it is a more accurate measurement than simply 25 centimetres.

Be careful with whole numbers. A crowd at a football match of 74 000 to the nearest thousand only has 2 significant figures. However, when you say that there are 90° in a right angle, there are 2 significant figures as the number is exactly 90. In other words, when there are trailing zeros in a whole number, it depends on whether the number has been rounded or is exact.

DON'T FORGET

Zeros which are used to indicate the position of the decimal point are not significant. The rules for rounding remain the same, that is *round up if 5 or more* or *do not round up if 4 or less.*

EXAMPLE

Round 33 528·746 to
2 significant figures.

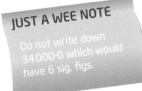

JUST A WEE NOTE

Do not write down 34 000·0 which would have 6 sig. figs.

SOLUTION

- 33 528·746 has 8 sig. figs.
- 33 528·746 lies between 33 000 and 34 000 (each of which has 2 sig. figs)
- look at the value of the third significant figure (33 **5**28·746), that is 5
- As the third significant figure is 5, round up to 34 000

EXAMPLE

A cylinder has height 25 centimetres and diameter 12 centimetres.

Calculate its volume. Give your answer correct to two significant figures.

(Volume of a cylinder = $\pi r^2 h$)

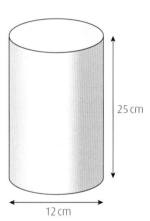

25 cm

12 cm

SOLUTION

$V = \pi r^2 h = \pi \times 6^2 \times 25 = 2827 \cdot 433388$

Hence volume of cylinder = 2800 cm³ correct to 2 sig. figs

DON'T FORGET

Remember to halve the diameter in order to use the radius in the formula. You can calculate $\pi \times 6^2 \times 25$ either by using the x^2 key on your calculator for 6^2 or by keying in $\pi \times 6 \times 6 \times 25$. Note that on some occasions when you are asked to find the volume of a cylinder, you may be told the height and the area of the base. If so, use the formula $V = Ah$.

COURSE IDEA

The topic of significant figures can be quite confusing. To help you understand things better, go online to find a significant figures calculator. You can practise here by trying different numbers and finding the solution.

THINGS TO DO AND THINK ABOUT

1. Round 6·0092 to two decimal places.

2. Round 5·98 to one decimal place.

3. Round 17·4999 to two significant figures.

4. Round 6786 to three significant figures.

5. Calculate the area of a circle of radius 13·5 centimetres.
   Give your answer correct to two significant figures.

DECIMALS

WORKING WITHOUT A CALCULATOR

After studying this section, without a calculator you should be able to

- add and subtract decimals
- multiply and divide decimals by a single digit whole number or by 10 or 100 or 1000
- write decimals as fractions.

Cover up the solutions to the next group of examples and do not use a calculator.

EXAMPLE

Work out 6·2 – 2·57

METHOD

Line up the numbers carefully so that the decimal points are in line:

```
 6·2
 2·57
```

Fill in any gaps with zeros so the numbers are the same length:

```
 6·20
 2·57
```

Then subtract the numbers as normal.
Insert a decimal point in the solution.

SOLUTION

3·63

DON'T FORGET

If you are adding and subtracting decimals without a calculator, line up the numbers so that the decimal points are in line. If there is a whole number involved, for example 5, you should write 5·0. Fill in any gaps with zeros so that the numbers are the same length. Remember to put the decimal point in the solution.

EXAMPLE

(a) How much will it cost to buy six paperback novels if they each cost £7·89?
(b) Divide £15 by 7. Round your answer to the nearest penny.

SOLUTION

(a) £47·34
(b) £2·14

Working for part (b)

```
        2 ·  1  4  2
  7 | 1 5 · ¹0 ³0 ²0
```

DON'T FORGET

If you are dividing a sum of money by a whole number and are asked to round your answer to the nearest penny, then this is the same as rounding to two decimal places. In the above example, three zeros were added after the decimal point as the third number after the point in the division tells you how to round the answer.

EXAMPLE

To multiply a decimal by 10 or 100 or 1000, move the decimal point to the right, one place for each zero. To divide a decimal by 10 or 100 or 1000, move the decimal point to the left, one place for each zero.

Work out

(a) 6·3 × 100
(b) 57·6 ÷ 10

SOLUTION

(a) 630
(b) 5·76

WRITING A DECIMAL AS A FRACTION

EXAMPLE

Write the following decimals as fractions:

(a) 0·9
(b) 0·17
(c) 0·023.

SOLUTION

(a) $\frac{9}{10}$
(b) $\frac{17}{100}$
(c) $\frac{23}{1000}$

 DON'T FORGET

The first number after the decimal point represents *tenths,* the second number after the decimal point represents *hundredths* and the third number after the decimal point represents *thousandths.*

USING A CALCULATOR

For more difficult examples, you will be allowed to use a calculator.

EXAMPLE

Rovers are organising coaches to take 926 supporters to an away match. A company offers them the following types of coach.

TYPE	COST
15 seat minibus	£180
45 seat deluxe coach	£325
75 seat double decker	£480

Rovers want to keep the cost as low as possible. Calculate the lowest possible cost of hiring the coaches?

SOLUTION

The larger the coach the more economical (better value) it becomes.

As the 75 seat double decker is the best value, use as many of this type of coach as possible.

Number of double decker coaches = 926 ÷ 75 = 12·34666667

This means that 12 double decker coaches could be filled with 12 × 75 = 900 supporters on them. There will still be 926 – 900 = 26 supporters remaining. They could all fit on one deluxe coach (which holds 45) or two minibuses (which would hold 2 × 15 = 30). It would cost 2 × £180 = £360 for two minibuses, therefore it is cheaper to hire the deluxe coach.

Therefore Rovers should hire 12 double decker coaches and one deluxe coach.

The cost = 12 × £480 + 1 × £325 = £5760 + £325 = £6085

COURSE IDEA

Investigate how to multiply and divide decimals by multiples of 10, 100 and 1000 without a calculator.

For example,
6·25 × 300 = 6·25 × 3 × 100 = 18·75 × 100 = 1875
and 9·6 ÷ 40 = 9·6 ÷ 4 ÷ 10 = 2·4 ÷ 10 = 0·24.

THINGS TO DO AND THINK ABOUT

Do not use a calculator for these examples.

1. Work out
   (a) 6·5 – 1·23

   (b) 7·58 × 9

   (c) 5·627 × 100

2. Jagtar buys a newspaper for £1·25 and a magazine for £3·49 at the newsagents. He pays for these items with a £10 note. How much change will he receive?

FRACTIONS AND RATIOS

SIMPLIFYING A FRACTION

To express a fraction in its simplest form, you must divide the **numerator** (the top number) and the **denominator** (the bottom number) in the fraction by their highest common factor (HCF). For example, if you wish to simplify $\frac{10}{25}$, then divide 10 and 25 by 5 (the HCF of 10 and 25) leading to $\frac{2}{5}$. Note that an alternative way of writing $\frac{2}{5}$ is 2/5. If you are asked to simplify a fraction with large numbers, you can do this in several steps. Start by dividing the numerator and denominator by small numbers, for example 2 or 3 until it becomes more manageable.

EXAMPLE

Simplify $\frac{72}{108}$

SOLUTION

Start by dividing by 2 until the numbers are smaller.

$$\frac{72}{108} = \frac{36}{54} = \frac{18}{27} = \frac{2}{3}$$

USING FRACTIONS

A very important use of fractions occurs when you are asked to work out a fraction of a number, shape or measurement. This occurred in the section on Scale factors when we had to enlarge a shape using a scale factor of $\frac{3}{2}$. Do not use a calculator for the following example.

EXAMPLE

John earns £112 from his part-time job. He gives $\frac{3}{7}$ of his earnings to his parents to help with household expenses. How much does he give to his parents?

SOLUTION

Work out $\frac{1}{7}$ of £112 by dividing £112 by 7 then work out $\frac{3}{7}$ of £112 by multiplying the answer by 3.

$\frac{3}{7}$ of £112 = £112 ÷ 7 × 3 = £16 × 3 = £48

CHANGING A FRACTION TO A DECIMAL

We can convert any fraction to a decimal fraction by dividing the numerator by the denominator. For example, check using a calculator that $\frac{3}{4}$ = 3 ÷ 4 = 0·75. Some fractions will give a decimal solution which goes on forever, for example $\frac{1}{3}$ = 1 ÷ 3 = 0·33333333... and so on. When this happens, we are likely to be asked to round the answer to a given number of decimal places.

EXAMPLE

Express $\frac{3}{7}$ as a decimal fraction. Give your answer correct to three decimal places.

SOLUTION

$\frac{3}{7}$ = 3 ÷ 7 = 0·4285714286 = 0·429 (correct to three decimal places)

Note that if no calculator is allowed for the above question, you would need to divide 3·0000 by 7.

COURSE IDEA

There is a useful key on most scientific calculators for dealing with fractions. Look for a key with a^{b}/c or $\frac{\blacksquare}{\square}$. This can be used for many calculations. To simplify $\frac{72}{108}$ (see above), key in 72 a^{b}/c 108 = and you should get $\frac{2}{3}$. Many other calculations involving fractions can be solved quickly using this key, but do not become too reliant on it as fractions are often assessed without the use of a calculator.

RATIOS

A ratio compares two different things. In the diagram we can see eight red squares and 20 yellow squares. We say that the ratio of red squares to yellow squares is 8 to 20. This can be written as a fraction, 8/20 or more usually written as 8:20. The ratio of yellow squares to red squares is 20:8. We can multiply or divide a ratio by the same number. If we are asked to express a ratio in its simplest form, we should divide the numbers in the ratio by their HCF (in the same way as we simplify fractions). If we divide 8 and 20 by 4, we find that the simplest form of the ratio 8:20 is 2:5. This means that for every 2 red squares there are 5 yellow squares. You can see that every row has 2 red squares and 5 yellow squares.

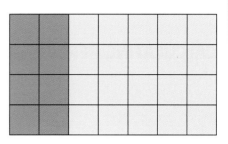

EXAMPLE

Express the ratio 24:32 in its simplest form.

SOLUTION

3:4 (Divide 24 and 32 by 8 which is the HCF of 24 and 32)

EXAMPLE

A cake mixture requires two eggs and three tablespoons of flour. Anna is going to make a large number of cakes. She uses 18 tablespoons of flour. How many eggs should she use?

SOLUTION

Compare quantities of flour. As 18 is $18 \div 3 = 6$ times bigger than 3, she will require $6 \times 2 = 12$ eggs.

DIVIDING A QUANTITY IN A GIVEN RATIO

EXAMPLE

There are 288 pupils on a school roll.
The ratio of boys to girls is 5:4.
How many girls are on the school roll?

SOLUTION

The ratio 5:4 means that on the roll there are five boys for every four girls.
So always start this type of question by adding the numbers in the ratio giving $5 + 4 = 9$. Therefore $\frac{5}{9}$ of the pupils are boys and $\frac{4}{9}$ of the pupils are girls.

Hence the number of girls = $\frac{4}{9}$ of 288 = $288 \div 9 \times 4 = 128$.

Check that $\frac{5}{9}$ of 288 = $288 \div 9 \times 5 = 160$ and that $128 + 160 = 288$.

DON'T FORGET

Remember that to express a fraction or a ratio in its simplest form you should divide the numbers in the ratio or fraction by their HCF. Practise examples of the type given in the section on 'Dividing a quantity in a given ratio' as this if often asked in assessments.

 THINGS TO DO AND THINK ABOUT

1. Express $\frac{48}{80}$ as a fraction in its simplest form.

2. Express $\frac{5}{8}$ as a decimal fraction.

3. Derek has 96 films in his DVD collection. If $\frac{3}{8}$ of them are war films, how many war films does he have in his collection?

4. In a box of chocolates the ratio of milk to plain chocolates is 3:2. If the box contains 40 chocolates, how many of them are milk chocolates?

PERCENTAGES

CALCULATING PERCENTAGES

In order to help calculate a percentage of a quantity, you should be able to convert a percentage to a fraction or a decimal fraction.

EXAMPLE

Convert 45% to

(a) a fraction in its simplest form
(b) a decimal fraction.

SOLUTION

(a) $45\% = \frac{45}{100} = \frac{9}{20}$ (divide numerator and denominator by 5)
(b) $45\% = 45 \div 100 = 0 \cdot 45$

There are certain commonly used percentages for which you should know the equivalent fraction. These are $1\% = \frac{1}{100}$, $10\% = \frac{1}{10}$, $20\% = \frac{1}{5}$, $25\% = \frac{1}{4}$, $33\frac{1}{3}\% = \frac{1}{3}$, $50\% = \frac{1}{2}$, $66\frac{2}{3}\% = \frac{2}{3}$ and $75\% = \frac{3}{4}$

EXAMPLE

Work out 25% of £80 without a calculator.

SOLUTION

25% of £80 $= \frac{1}{4}$ of £80 = £80 $\div$ 4 = £20

EXAMPLE

Calculate 35% of £120

(a) with a calculator and
(b) without a calculator.

SOLUTION

(a) 35% of £120 = $0 \cdot 35 \times$ £120 = £42
(b) 10% of £120 = £12, hence 30% of £120 = 3 × £12 = £36 and
    5% of £120 = £12 $\div$ 2 = £6
    Hence 35% of £120 = £36 + £6 = £42

CONVERTING FRACTIONS TO PERCENTAGES

To convert a fraction to a percentage, multiply the fraction by 100.

EXAMPLE

(a) Convert $\frac{4}{5}$ to a percentage.
(b) Alistair buys a painting. It costs him £300. He later sells it for £375. Express his profit as a percentage of the cost price.

SOLUTION

(a) $\frac{4}{5} = \frac{4}{5} \times 100\% = 4 \div 5 \times 100\% = 80\%$
    Note that without a calculator, you should divide 100 by 5 then multiply by 4.
(b) Actual profit = £375 – £300 = £75
    Profit as a percentage of cost price $= \frac{75}{300} \times 100\%$
    $= 75 \div 300 \times 100\%$
    $= 25\%$

THE EQUIVALENCE OF FRACTIONS, DECIMALS AND PERCENTAGES

We have studied how to convert between fractions, decimals and percentages. We know, for example, that $\frac{1}{2} = 0.5 = 50\%$. We can say that the fraction, decimal and percentage are equal or equivalent. In certain situations it may be that one of the three versions is most suited.

EXAMPLE

List the following in ascending order (from smallest to largest):

$0.8, \quad \frac{17}{20}, \quad 78\%, \quad 0.76, \quad \frac{21}{25}$

 DON'T FORGET

If you have to compare the size of fractions, decimals and percentages, you should convert them to the same form. As it can be difficult to compare the size of certain fractions, it is usually simplest to convert to decimal form.

SOLUTION

We must make sure that the five amounts are all in the same form in order to compare them, but which form – fraction, decimal or percentage? Good advice is **not** to convert all the amounts to fractions as it can be difficult and time-consuming to compare fractions. If you are allowed to use a calculator, the simplest method is to convert to decimal form.

$0.8 = 0.80, \quad \frac{17}{20} = 17 \div 20 = 0.85, \quad 78\% = 0.78, \quad 0.76, \quad \frac{21}{25} = 21 \div 25 = 0.84$

Hence the correct ascending order is $0.76, 78\%, 0.8, \frac{21}{25}, \frac{17}{20}$

EXAMPLE

Martina has sat three tests as part of her mathematics assessment. Her marks are listed below.

Unit One: 45 out of 61   Unit Two: 57 out of 80   Unit Three: 37 out of 55

In which test did Martina perform best. Use your calculations to justify your answer.

SOLUTION

$45 \div 61 = 0.737704918, \quad 57 \div 80 = 0.7125, \quad 37 \div 55 = 0.6727272727$

As the largest decimal is 0.737704918, Martina performed best in the Unit One test.

COURSE IDEA

Percentages occur often in the news. You have probably heard the word inflation mentioned. Try to find out about inflation. It is usually given as a percentage and shows how the price of goods and services has changed over a period of time.

 THINGS TO DO AND THINK ABOUT

Do not use a calculator for Questions 1 – 3.

1. Convert 60% to a fraction in its simplest form.

2. Work out 20% of 95 kilograms.

3. There are 500 cars in a car park. 45% of the cars are silver. How many silver cars are in the car park?

4. Alison has bought 25 packets of crisps for a party. 16 of the packets are ready salted flavour. What percentage of the packets is ready salted flavour?

5. In her exams, Maria has scored the following marks:
   Geography: 28 out of 40,   French: 17 out of 25,   English: 46 out of 64

   In which subject did Maria perform best? Use your calculations to justify your answer.

SAVING AND SPENDING

PERCENTAGE INCREASE AND DECREASE

When dealing with money, there are many areas where a percentage increase or decrease can arise, for example, a pay increase (often given in line with inflation), **value added tax** or VAT for short (which is added to the cost of certain goods) or a **discount** (which reduces the price of an item in a sale). Some employees who sell goods for a living are paid **commission** (a percentage of their sales) in addition to a basic salary.

EXAMPLE

A laptop costs £350 plus VAT. If VAT is charged at 20%, find the total cost of the laptop.

SOLUTION

VAT = 20% of £350 = 0·20 × £350 = £70

So the total cost is £350 + £70 = £420

EXAMPLE

A store offers a discount of $33\frac{1}{3}$% on the marked price of all its goods. Find the sale price of a vacuum cleaner whose price is marked at £159.

SOLUTION

Discount = $33\frac{1}{3}$% of £159 = $\frac{1}{3}$ of £159 = £159 ÷ 3 = £53

So the sale price is £159 − £53 = £106

SAVING MONEY

A sensible way of saving money is to deposit a sum of money in a bank where it will earn **interest**. Interest is given as a percentage per annum. Per annum means per year. Interest can be calculated for a period of time less than a year, for example, the interest for 6 months would be $\frac{6}{12}$, that is half the interest for a full year, as there are 12 months in a year.

EXAMPLE

Wendy deposits £2400 in the bank. The rate of interest is 2% per annum. How much interest will Wendy earn in 4 months?

SOLUTION

Interest for a full year = 2% of £2400 = 0·02 × £2400 = £48

Interest for 4 months = $\frac{4}{12}$ of £48 = £48 ÷ 12 × 4 = £16

HIRE PURCHASE

It is possible to buy items on **hire purchase**, or HP for short. This is one way of buying expensive items. You start by paying a small part of the cost, called the deposit. Once this is paid, you can take the item home. The rest is paid by monthly repayments, sometimes called monthly instalments.

EXAMPLE

Dorothy buys a washing machine on HP. She pays a deposit of £55 followed by 24 monthly repayments of £11·99. If the cash price of the washing machine is £275, find the extra cost of hire purchase.

SOLUTION

Cost of repayments = 24 × £11·99 = £287·76

Total HP cost = deposit + repayments = £55 + £287·76 = £342·76

Extra cost of HP = £342·76 – £275 = £67·76

DON'T FORGET

To convert British currency (£) to foreign currency, multiply the number of pounds by the exchange rate. To convert foreign currency to British currency (£), divide the amount of foreign currency by the exchange rate.

FOREIGN EXCHANGE

When travelling abroad, it is necessary to convert British currency (pounds sterling) into foreign currency. The amount of foreign currency you receive depends on the exchange rate. This changes regularly and can be found online and in newspapers daily.

EXAMPLE

The exchange rate for pounds to euros is £1 = 1·22 euros.

(a) Convert £500 to euros.
(b) Convert 800 euros to pounds. Give your answer to the nearest penny.

SOLUTION

(a) £500 = 500 × 1·22 euros = 610 euros
(b) 800 euros = 800 ÷ 1·22 pounds
    = £655·7377049
    = £655·74 (to the nearest penny)

COURSE IDEA

The euro is used in many European countries including Germany, France, Spain and Italy. The collective name for the countries using the euro is the Eurozone. Find out what countries are in the Eurozone and investigate the currencies of other countries around the world.

THINGS TO DO AND THINK ABOUT

1. Craig's annual salary is £16 200. Find his new annual salary if he is given a pay rise of 1%.

2. Isobel is a sales representative who is paid commission of 5% on everything she sells. One month she sells goods worth £24 000. What is her commission for that month?

3. Mary deposits £800 in the bank. The rate of interest is 3% per annum. How much interest will Mary earn in 6 months?

4. The exchange rate for pounds to US dollars ($) is £1 = $1·66.
   (a) Convert £400 to US dollars.
   (b) Convert $50 to pounds. Give your answer to the nearest penny.

5. Neil buys a car on hire purchase by paying a deposit of £2750 followed by 24 instalments of £475. Find the total hire purchase price of the car.

MEASUREMENT

MEASURING

Check that you can measure to a satisfactory degree of accuracy with a ruler and protractor.

EXAMPLE

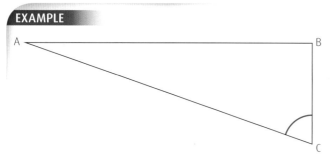

Measure

(a) the length of side AC
(b) the size of angle ACB in triangle ABC shown above.

DON'T FORGET

Although exact measurements are not always possible, you should try to be as accurate as you can when using a ruler and protractor.

SOLUTION

(a) 7·6 cm
(b) 71°

SOME UNITS OF MEASUREMENT

Make sure you remember the connections between the basic units of measurement in the metric system.

- Length: 1 cm = 10 mm   1 m = 100 cm   1 km = 1000 m
- Volume: 1 ml = 1 cm³   1 l = 1000 ml
- Mass: 1 g = 1000 mg   1 kg = 1000 g

EXAMPLE

Convert

(a) 560 centimetres to metres
(b) 0·075 grams to milligrams

SOLUTION

(a) 560 cm = (560 ÷ 100) m = 5·6 m
(b) 0·075 g = (0·075 × 1000) mg = 75 mg

DON'T FORGET

The prefixes which come *before* measurements have special meanings, for example k for kilo means 1000. Other useful prefixes are m for milli meaning $\frac{1}{1000}$ as in a milligram is $\frac{1}{1000}$ th of a gram, c for centi meaning $\frac{1}{100}$ and d for deci meaning $\frac{1}{10}$. Therefore a volume of 75 cl means 75 centilitres and is $\frac{75}{100}$ or 0·75 of a litre. For very large measurements, the prefix M for mega means 1 000 000 (1 million), for example a measurement of 1 MW means 1 megawatt = 1 000 000 watts where the watt is a measurement of power.

READING SCALES

In every household, there are a variety of measuring instruments which require us to read scales. In the kitchen, there are clocks and timers, scales for weighing and jugs for measuring liquids. A boiler may have timers and temperature settings. A barometer measures atmospheric pressure and may include a thermometer. Many bathrooms have a set of scales. Enter a car and you will see a milometer, a speedometer, a petrol gauge and so on. Therefore it is vital that you can read the scales on such instruments.

EXAMPLE

Write down the reading shown on each of the following scales.

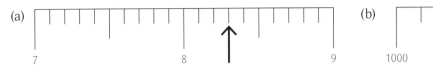

(a)

7   8   9

(b)

1000   1010

SOLUTION

(a) 8·3

(b) 1004

METHOD

In part (a), there are 10 divisions (spaces) between 8 and 9 so each division represents 0·1. In part (b) there are 5 divisions between 1000 and 1010. As 1010 – 1000 = 10 and 10 ÷ 5 = 2, each division represents 2.

EXAMPLE

1·5 litres

Orange juice has been poured into a measuring jug. How much extra orange juice should be added to fill the jug to the 1·5 litre mark?

EXAMPLE

James looks at the petrol gauge in his car. The purple arrow indicates how much petrol is left in the tank. The red area is a warning that petrol is running low.

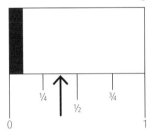

¼   ½   ¾

0   1

James decides to fill up his tank. His tank holds 48 litres when full. How many litres should James add to his tank in order to fill it up?

SOLUTION

There are already 0·6 litres of orange juice in the jug.

The amount to be added = (1·5 – 0·6) litres = 0·9 litres.

SOLUTION

The arrow indicates that the number of litres in the tank is midway between $\frac{1}{4}$ of 48 = 12 and $\frac{1}{2}$ of 48 = 24. Midway between 12 and 24 is 18 as the mean of 12 and 24 = (12 + 24) ÷ 2 = 18. Therefore there are 18 litres in the tank. So he should add (48 – 18) litres = 30 litres to the tank.

COURSE IDEA

Check out the recipe for a popular dish such as spaghetti Bolognese. You will find that plenty of measuring is required from teaspoons and tablespoons to measuring weight and volume. Do you think you could prepare this meal?

THINGS TO DO AND THINK ABOUT

1. Convert 135 millilitres to litres.

2. Convert 1500 centilitres to litres.

3. Write down the reading shown on the scale.

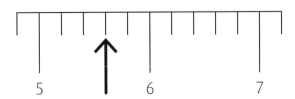

5   6   7

PROPORTION

WHAT IS RATE IN MATHEMATICS?

Before we look at proportion, it will be helpful to find out what we mean by the word *rate* in mathematics. A rate is a special type of ratio in which two measurements are related to one another. In a rate, the measurements are in different units, for example miles per gallon. Rates are often recognised by the word 'per' between the units involved where per means 'for every'. Therefore 40 miles per gallon means that a car can drive 40 miles for every gallon of petrol in the tank. Other examples of rates are miles per hour (mph), kilometres per hour (km/h), words per minute (related to the speed of a typist) and heart beats per minute.

EXAMPLE

The scale below shows speed in miles per hour and kilometres per hour.

speed in kilometres per hour

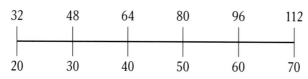

| 32 | 48 | 64 | 80 | 96 | 112 |

| 20 | 30 | 40 | 50 | 60 | 70 |

speed in miles per hour

Use the scale to change 45 miles per hour to kilometres per hour.

SOLUTION

45 mph is midway between 40 and 50 leading to the solution being midway between 64 and 80 km/h. The mean of 64 and 80 = (64 + 80) ÷ 2 = 144 ÷ 2 = 72, hence the solution is 72 km/h.

DIRECT PROPORTION

Two quantities are said to be in **direct proportion** if, as one quantity increases or decreases, the other increases or decreases *at the same rate*. For example, if a car can drive 40 miles per gallon of petrol, then it could drive 80 miles on 2 gallons or 20 miles on half a gallon.

EXAMPLE

Susan can type 500 words in 4 minutes. Complete the table below.

Number of minutes	1	2	3	4	5	6
Number of words				500		

SOLUTION

Number of minutes	1	2	3	4	5	6
Number of words	125	250	375	500	625	750

METHOD

Divide 500 by 4 to find the rate of 125 words per minute. This is called the *unit rate* as it tells you how many words can be typed in *1 minute*. It can be used to find the solutions to any other number of minutes by multiplication. Note that the ratio of the number of words to the number of minutes (125:1, 250:2, 375:3, 500:4 and so on) is constant because all these ratios simplify to 125:1. We can see that the number of minutes and the number of words are in direct proportion.

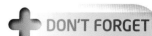

EXAMPLE

Bob drives 540 kilometres on a motorway in 5 hours. Travelling at the same speed, how far could he travel in 3 hours?

SOLUTION

Number of hours		Number of kilometres
5	→	540
1	→	540 ÷ 5 = 108
3	→	108 × 3 = 324

Hence Bob could travel 324 km in 3 hours.

If, when you are finding the unit rate, the answer to the division sum is a decimal fraction which goes on forever, keep calm as this can happen. Do not round the decimal fraction. See the next example.

DON'T FORGET

Working should be set out as shown in two columns with the right-hand column being the quantity you are asked to find. Then start by finding the unit rate. Note that common sense should tell you that the answer must be *less than* 540. In fact the number 540 has been reduced in the ratio $\frac{3}{5}$. Note that the smaller number is on the numerator of the fraction. It is useful to think whether your answer makes sense or not in proportion examples as some students would mistakenly calculate 540 ÷ 3 × 5 leading to 900 which clearly does not make sense.

EXAMPLE

Chang earns £202·50 for working 30 hours in a supermarket.
How many hours would he need to work to earn £243?

SOLUTION

Earnings		Number of hours
£202·50	→	30
£1	→	30 ÷ 202·50
£243	→	30 ÷ 202·50 × 243 = 36

Hence Chang would earn £243 in 36 hours.

JUST A WEE NOTE

The answer to the intermediate step (30 ÷ 202·50) appears as 0·1481481481 on a calculator. If you get an answer like this, you do not need to write it down. Simply keep it on your screen (without rounding) then multiply by 243, the final step leading to the correct answer of 36 hours. Check that your answer makes sense.

COURSE IDEA

Consider the following situation. A man employs six men to complete a building job in 20 days. How many days would it take to complete the job if he employed eight men instead? Can you work out the answer? If you think about it, by employing extra men, the job will be completed in a shorter time. In other words, as one quantity increases (the number of men), the other quantity decreases (the number of days). This is an example of *inverse* proportion. It is useful to have a think about examples like this. Well done if you worked out the solution to the problem which is 15 days.

THINGS TO DO AND THINK ABOUT

1. Pedro can buy 60 euros for £50. How many euros can he buy for £65?

2. 100 grams of chicken contains 23·5 grams of protein. How many grams of protein are there in a chicken weighing 2 kilograms?

NEGATIVE NUMBERS

COORDINATES

An **integer** is a member of a set of numbers containing positive whole numbers, negative whole numbers and zero. The set of integers can be written as {...–3, –2, –1, 0, 1, 2, 3...}. Integers appear on the *x*- and *y*-axes on a coordinate grid.

EXAMPLE

Plot the points A (–4, –2), B (–1, 4) and C (5, 4). Plot a fourth point D so that ABCD is a parallelogram. Write down the coordinates of D.

SOLUTION

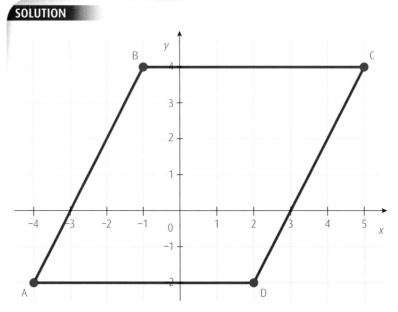

Point D has coordinates (2, –2).

ADDING INTEGERS

It is helpful when adding (and subtracting) integers to use a number line. The number line shown below can be extended as required.

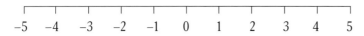

To add two integers, find the first number on the number line, then

- count to the *right* if you are adding a *positive* integer, for example –3 + 5 = 2
- count to the *left* if you are adding a *negative* integer, for example –3 + (–1) = –4

EXAMPLE

The temperature is –5°C at 2am one morning. By 10am the same morning the temperature has risen by 9°C. What is the temperature at 10am?

SOLUTION

–5 + 9 = 4

Hence at 10am the temperature is 4°C.

SUBTRACTING INTEGERS

To subtract two integers, find the first number on the number line, then

- count to the *left* if you are subtracting a *positive* integer, for example –3 – 2 = –5
- count to the *right* if you are subtracting a *negative* integer, for example –3 – (–8) = 5

JUST A WEE NOTE

Note that when you subtract a negative integer it is the same as adding the integer, for example
–3 – (–8) = –3 + 8 = 5

EXAMPLE

The temperature cools down from 4°C to –3°C one evening. By how many degrees did the temperature drop?

SOLUTION

$4 - (-3) = 7$

Hence the temperature dropped by 7°C.

DON'T FORGET

When you are adding and subtracting integers, it is a good idea to draw a number line to help. Learn the direction to move along the number line for addition and go the opposite direction for subtraction.

MULTIPLYING AND DIVIDING INTEGERS

The rules for multiplying and dividing integers are similar and are summarised in the tables below.

MULTIPLICATION
+ × + = +
+ × − = −
− × + = −
− × − = +

DIVISION
+ ÷ + = +
+ ÷ − = −
− ÷ + = −
− ÷ − = +

EXAMPLE

Work out

(a) $6 \times (-3)$
(b) -5×7
(c) $-30 \div (-10)$
(d) $-10 \div 2$

SOLUTION

(a) –18
(b) –35
(c) 3
(d) –5

DON'T FORGET

When you multiply two negative numbers together, the answer is positive. When you divide a negative number by another negative number, the answer is positive.

COURSE IDEA

Carrying out the four operations on integers without a calculator can be confusing. However, good knowledge of negative numbers is essential if you are to make progress in mathematics as they occur not only in examples on coordinates and temperatures, but also in equations. It is recommended that you practise some exercises of straightforward examples on the four operations until you have mastered the processes. You can also attempt the examples using a calculator. You may have to use keys such as +/- or (–).

THINGS TO DO AND THINK ABOUT

1. Work out
   (a) $-5 + 11$ (b) $7 + (-3)$ (c) $-4 + 4$ (d) $-7 - 2$ (e) $1 - (-5)$

2. Work out
   (a) $7 \times (-3)$ (b) -8×0 (c) $-1 \times (-1)$ (d) $-12 \div 3$

3. In a quiz, Stephen scored 12 points and Alan scored –5 points.
   How many points more than Alan did Stephen score?

TIMETABLES

TIME INTERVALS

When calculating a time interval, it is a good idea to count on your fingers rather than rely on a calculator. Suppose for example that you were asked to calculate how many days there were from 25 March to 4 May inclusive.

Count 25, 26, 27, 28, 29, 30 and 31 (7 days left in March), then add 30 days for the whole of April, then count 1, 2, 3 and 4 (first 4 days in May). You must include 25 March and 4 May as the word *inclusive* is used. Hence the answer is 7 + 30 + 4 = 41 days.

We shall now look at time intervals involving both the 12-hour clock and the 24-hour clock. Many students realise that they are finding the *difference* between two times when calculating a time interval, for example from 6.59pm to 7.02pm. They decide to subtract the times. This is correct but you should *never* use a calculator to subtract such times. To explain why, consider the interval from 6.59pm to 7.02pm. This is obviously 3 minutes yet some students will use a calculator, get 7·02 – 6·59 = 0·43 and think the answer is 43 minutes.

> *30 days has September,*
> *April, June and November.*
> *All the rest have 31*
> *Except February alone,*
> *Which has 28 days clear*
> *And 29 in each leap year.*

DON'T FORGET

Make sure you know how many days are in each month of a year.

EXAMPLE

Dan settles down to watch coverage of the cup final on television. Coverage starts at 1.25pm and finishes at 6.10pm. How long did the coverage of the cup final last?

SOLUTION

Start at 1.25pm and add on 1 hour at a time until you are less than an hour away from the finishing time, then add on the extra minutes.

1.25 → 2.25 → 3.25 → 4.25 → 5.25 = 4 hours   (1 hour per arrow)

5.25 → 6.00 = 35 minutes *and* 6.00 → 6.10 = 10 minutes

Hence the coverage lasted 4 hours + (35 + 10) minutes = 4 hours 45 minutes

---

EXAMPLE

Phil takes an overnight ferry from Hull to Rotterdam. The ferry departs from Hull at 2030 on Monday and arrives in Rotterdam at 0805 on Tuesday morning. How long did the ferry journey take?

SOLUTION

2030 → 2130 → 2230 → 2330 → 0030 → 0130 → 0230 → 0330 → 0430 → 0530 → 0630 → 0730 = 11 hours

0730 → 0800 = 30 minutes *and* 0800 → 0805 = 5 minutes

Hence the ferry journey takes 11 hours + (30 + 5) minutes = 11 hours 35 minutes.

DON'T FORGET

When calculating a time interval, count forward in hours (using your fingers as a check) and add on any extra minutes.

READING TIMETABLES

We all have to read timetables at some time when travelling on buses, trains, ferries and aeroplanes. Look at part of the train timetable from Glasgow Queen Street to Dundee shown below.

Glasgow Queen Street	d	0556	0741	0806	0841	0908
Larbert	d	0616	0752	0820	0851	0920
Stirling	d	0625	0809	0836	0908	0935
Dunblane	d	0631	0815	0844	✳	0941
Gleneagles	d	0643	✳	0856	✳	0953
Perth	a	0659	0841	0912	0939	1009
Perth	d	0700	0842	0915	0940	1015
Invergowrie	d	✳	✳	0939	✳	1034
Dundee	a	0722	0903	0944	1004	1043

Timetables are always given in 24-hour clock times. The vertical columns show the times of *five morning trains* from Glasgow Queen Street to Dundee. The letters *d* and *a* in the first column stand for depart and arrive, for example the first train shown arrives in Perth at 0659 and departs 1 minute later at 0700. You will see a small symbol (✳) opposite certain stations. This means that the train does not stop there, for example the first train shown does not stop at Invergowrie. Use the timetable for the next example.

Sandi has to travel by train from Dunblane to Dundee for an interview at 10.40am. Suggest which train Sandi should take from Dunblane. Justify your answer.

10.40am = 1040 as a 24-hour clock time. Sandi must arrive in Dundee well before 1040. Arriving in Dundee at 1004 would probably be OK, but if we check the train arriving in Dundee at 1004 we see that it does not stop at Dunblane. Therefore she should get the earlier train leaving Dunblane at 0844.

COURSE IDEA

Each day people miss trains, buses, ferries and even aeroplanes because they have made an error with the 24-hour clock or in reading a timetable. At best this is inconvenient, sometimes it is disastrous as they may be late for an interview or miss a holiday. To avoid such things happening, you should pay very close attention to timetables, including the small print at the bottom as this may indicate that some trains or buses only operate on certain days. Timetables appear online or are available at bus and train stations. Check out local timetables in the area where you live and set about understanding them thoroughly.

Arthur has calculated that a car journey should take him 3 hours 45 minutes. If he has to arrive at his destination at 11.20am, when should he set out?

Work backwards from 11.20:   11.20 → 10.20 → 9.20 → 8.20 = 3 hours.

Now work backwards 45 minutes from 8.20:   8.20 → 8.00 = 20 minutes and, as 20 + 25 = 45, 8.00 → 7.35 = 25 minutes, hence he should set out at 7.35am. You should check your answer by adding 3 hours 45 minutes onto 7.35am.

THINGS TO DO AND THINK ABOUT

1. A car trip starts at 1325 and ends at 1803. How long did the trip last?

2. Kenneth's flight takes off at 2236. The captain announces to the passengers that the estimated flight time is 3 hours 45 minutes. What is the estimated arrival time of Kenneth's flight?

DISTANCE, SPEED AND TIME

DISTANCE –TIME GRAPHS

Suppose the Murray family leave home at 9.00am and drive 40 miles to the seaside for a day out. They arrive at 10.00am, stay for 5 hours and set off for home again at 3.00pm. Because of traffic delays, the return journey takes 2 hours. The whole trip can be shown on a distance–time graph.

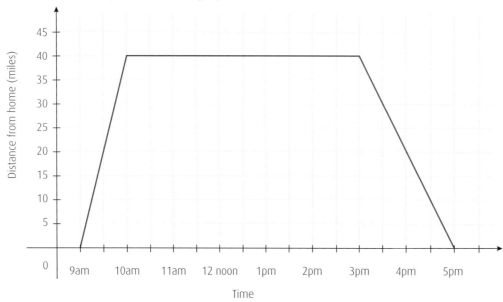

Note that the horizontal line represents times when the family are not travelling. The upward slope (positive gradient) represents the journey to the seaside and the downward slope (negative gradient) represents the journey home. Where the slope is steeper, the speed is greater. We can see that the journey to the seaside was completed at a greater speed (40 miles per hour) than the journey home (20 miles per hour).

DON'T FORGET

A distance–time graph is a visual way of showing a journey with a horizontal time axis and a vertical distance axis.

CALCULATING DISTANCE

The distance travelled by a moving object depends on two things – the speed of the object and the time taken. Common units of speed are miles per hour (mph), kilometres per hour (km/h) and metres per second (m/s). Notice that these are all examples of a rate. If a car travels 160 kilometres in 2 hours, we would say that its *average speed* was 80 km/h which is the unit rate. When a car travels at this speed, it will travel 80 km in 1 hour, 160 km in 2 hours and 240 km in 3 hours, etc. We can see that speed and distance are two quantities in direct proportion. By multiplying the average speed by the number of hours (the time), we can calculate distance. This is given as a formula:

Distance = Speed × Time   or $D = ST$ for short

EXAMPLE

A train travels at an average speed of 72 kilometres per hour for 5 hours. What distance was travelled?

SOLUTION

$D = ST = 72 \times 5 = 360$

Hence the distance travelled was 360 km.

When you are calculating distance, make sure that the units of distance speed and time are consistent, for example kilometres, kilometres per hour and hours respectively. If minutes are included in the time, you may have to convert the minutes to a decimal fraction of an hour. To convert minutes to a fraction of an hour, you should divide by 60, for example 30 minutes = 30 ÷ 60 = 0·5 hours. This makes sense as 30 minutes is half an hour. We can show that 15 minutes (quarter of an hour) = 0·25 hours and 45 minutes (three quarters of an hour) = 0·75 hours.

EXAMPLE

Marco drives for 2 hours 45 minutes at a speed of 112 kilometres per hour. What distance does he travel?

SOLUTION

$D = ST = 112 \times 2$ hours 45 minutes
$= 112 \times 2{\cdot}75 = 308$

Hence Marco travels 308 km.

 DON'T FORGET

You should convert times in minutes to a decimal fraction of an hour when speeds are given in mph or km/h. This is done by dividing the number of minutes by 60, for example 2 hours 18 minutes = 2·3 hours as 18 ÷ 60 = 0·3.

EXAMPLE

Part of the train timetable from Glasgow Central to London Euston is shown below.

Glasgow Central	Depart 1540
Carlisle	Depart 1648
Penrith North Lakes	Depart 1703
Lancaster	Depart 1737
Preston	Depart 1758
Wigan North Western	Depart 1809
Warrington Bank Quay	Depart 1820
London Euston	Arrive 2012

(a) Calculate the journey time from Carlisle to London Euston?

(b) The train travels at an average speed of 141·5 kilometres per hour between Carlisle and London Euston. Calculate the distance between Carlisle and London Euston. Give your answer to the nearest kilometre.

SOLUTION

(a) 1648 → 1748 → 1848 → 1948 = 3 hours
1948 → 2000 = 12 minutes *and* 2000 → 2012 = 12 minutes

Hence journey time is 3 hours + (12 + 12) minutes
= 3 hours 24 minutes

(b) 24 minutes = 24 ÷ 60 = 0·4 hours, so 3 hours 24 minutes = 3·4 hours
$D = ST = 141{\cdot}5 \times 3{\cdot}4 = 481{\cdot}1$

Hence the distance between Carlisle and London Euston = 481 km (to the nearest kilometre).

COURSE IDEA

By changing the subject of the formula $D = ST$ to S and T, we can form two new formulae, namely $S = \frac{D}{T}$ and $T = \frac{D}{S}$. The first of these formulae can be used to calculate speed when you know distance and time. The second of these formulae can be used to calculate time when you know distance and speed. Suppose for example that a train travelled 400 kilometres in 2 hours 30 minutes, then we could find the speed of the train as follows: $S = \frac{D}{T} = \frac{400}{2{\cdot}5} = 160$, leading to a speed of 160 km/h. Investigate both formulae.

THINGS TO DO AND THINK ABOUT

1. A car drives at an average speed of 70 miles per hour for 3 hours 30 minutes. What distance did the car travel?

2. How far would a bus travel in 45 minutes at a steady speed of 56 kilometres per hour?

AREA AND PERIMETER

REAL-LIFE PROBLEMS

Problems on area and perimeter arise often in daily life, for example when decorating, wallpapering, laying carpets and erecting fences. We have already looked at area and perimeter in earlier sections but will now look at some extended problems on these topics.

EXAMPLE

Lewis and Manuel are going to play a game of tennis. They decide to warm up by jogging around the perimeter of the tennis court. The court is in the shape of a rectangle measuring 24 metres by 11 metres.

Lewis claims that if they jog around the perimeter of the tennis court 15 times then they will each have jogged more than 1 kilometre.

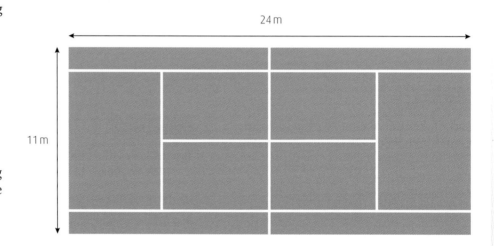

Is Lewis correct? Justify your answer by calculation.

SOLUTION

Perimeter of rectangle = (24 + 11 + 24 + 11) m = 70 m

Distance jogged = 15 × 70 m = 1050 m

Lewis is correct as 1 kilometre = 1000 m and 1050 > 1000.

EXAMPLE

Fiona is going to plant grass seed in part of her garden.
The area to be planted is shown below.

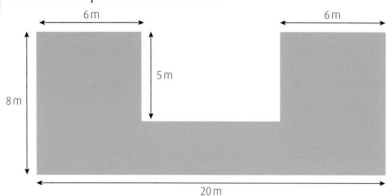

Fiona has bought five kilograms of grass seed for the lawn. She knows that 50 grams of grass seed are needed for each square metre of lawn.

Has Fiona bought enough grass seed? Justify your answer by calculation.

DON'T FORGET

A reminder that when you are asked to justify your answer by calculation, you must show all the calculations carried out in your working. You must also compare both numerical values or state the difference between them as well as reaching a conclusion. Check carefully all the working in the two previous examples.

SOLUTION

Find the area of the lawn by subtraction.

Note that the length of the 'missing' rectangle must be $(20 - 6 - 6)$ m = 8 m

Area of 'full' rectangle: $A = lb = 20 \times 8 = 160$

Area of 'missing' rectangle: $A = lb = 8 \times 5 = 40$

Area of lawn: $A = 160 - 40 = 120$, so Fiona has to cover 120 m² of lawn.

She will need 120×50 g = 6000 g = 6 kg of grass seed.

Hence Fiona has not bought enough grass seed. She should have bought 1 more kilogram.

DECORATING A ROOM

EXAMPLE

Paul has built a garden hut. It is in the shape of a cuboid. It is 4 metres long, 3·5 metres broad and 2·5 metres high. The door measures 2 metres by 1·2 metres and the window measures 1·2 metres by 0·5 metres.

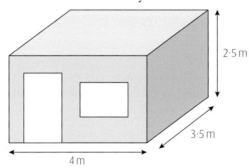

Paul has decided to paint the four walls inside the hut. He will not paint over the door or window. The paint he has chosen costs £2·10 per square metre of coverage. Calculate the total cost of painting the room.

SOLUTION

Area of four walls
$= (4 \times 2\cdot5) + (4 \times 2\cdot5) + (3\cdot5 \times 2\cdot5) + (3\cdot5 \times 2\cdot5)$
$= 37\cdot5$ m²

Area of door and window
$= (2 \times 1\cdot2) + (1\cdot2 \times 0\cdot5) = 3$ m²

Area to be painted $= 37\cdot5 - 3 = 34\cdot5$ m²

Cost of paint $= 34\cdot5 \times £2\cdot10 = £72\cdot45$

COURSE IDEA

It is convenient to use an online paint calculator and wallpaper calculator when decorating a room. They will tell you the amounts that you will need.

Check these online and find out what measurements are required for the online calculator.

THINGS TO DO AND THINK ABOUT

A room in the shape of a trapezium is to be fitted with a new carpet.

Find the cost of fitting the carpet if the carpet costs £4·99 per square metre.

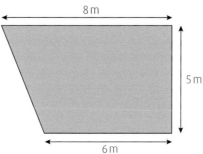

INFORMATION FROM TABLES AND GRAPHS

EXTRACTING AND INTERPRETING INFORMATION

You should be able to extract and interpret data from tables and graphs as well as make decisions based on the interpretation of data.

EXAMPLE

The Bell family is planning a seven-night holiday. They decide to check prices and temperatures at the Riviera in the south of France and are given a holiday brochure at their local travel agent.

Month	Hotel Cote d'Azur	Hotel Monaco	Hotel de la Plage
Jan–Mar	£688	£774	£869
Apr–May	£796	£880	£935
Jun–Jul	£975	£1020	£1140
Aug	£902	£972	£1047
Sep–Oct	£810	£933	£957
Nov–Dec	£715	£770	£880

The above table shows the price per person for a seven-night all inclusive holiday to three popular hotels in the area. Children under 5 years of age pay half-price.

The line graph below shows the annual temperature in degrees Celsius in the region.

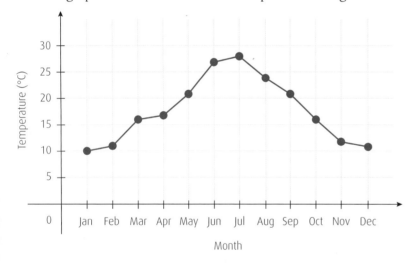

The family wish to go at a time when the temperature is between 20°C and 25°C. They want to go on the cheapest possible holiday during that time.

The family consists of Mr Bell, Mrs Bell and their daughter Kirsty aged 3. Which holiday should they choose and how much will it cost?

SOLUTION

By looking at the line graph in the temperature chart, we can see that only three months have temperatures between 20°C and 25°C – May, August and September. We should then look at the cost of each hotel in the table of prices during these three months and look for the cheapest price. This shows that the family should choose to go to the Hotel Cote d'Azur in May. This costs £796 per person but Kirsty will pay half-price as she is under 5 years of age.

Hence cost = £796 + £796 + $\frac{1}{2}$ of £796 = £796 + £796 + £398 = £1990

COURSE IDEA

Look at some holiday brochures, either from a travel agent or online. Try to understand the many different tables and graphs in the brochure. You may even get some good holiday ideas in the process.

STEM AND LEAF DIAGRAMS

A **stem and leaf diagram** presents data in a clear and ordered way.

If you are asked to draw a stem and leaf diagram, remember the following points.

- Each number in a data set is split into a stem and a leaf.
- The leaf is the last digit in each number, for example 3 in 23.
- The stem is the remaining digit(s) in each number, for example 2 in 23.
- The stems should be written to the left of a vertical line in the diagram in an ordered column.
- The leaves should be written in order to the right of the vertical line.
- The diagram should contain a title, a key and show the number of items in the data set.

EXAMPLE

The marks of a class of students in a numeracy test are listed below.

45 68 74 50 43 38 74 52 68 80

70 48 38 73 67 66 89 54 63 66

70 56 48 65 65 60 56 47 35 62

(a) Illustrate this data in an ordered stem and leaf diagram.

(b) Find the median mark.

SOLUTION

(a) NUMERACY TEST MARKS

3	5 8 8
4	3 5 7 8 8
5	0 2 4 6 6
6	0 2 3 5 5 6 6 7 8 8
7	0 0 3 4 4
8	0 9

$n = 30$ 3 | 5   represents 35

(b) As $n = 30$, position of median = $(30 + 1) \div 2 = 15 \cdot 5$.
This means the median is between the 15th and 16th values. As the numbers are ordered, the median is therefore the mean of 62 and 63 which is calculated as $(62 + 63) \div 2 = 125 \div 2 = 62 \cdot 5$
Hence the median mark = $62 \cdot 5$

 **DON'T FORGET**

Check that you can draw a stem and leaf diagram. Always remember to say what n is and include a key. Always check that you have included all the leaves by counting.

COURSE IDEA

A back-to-back stem and leaf diagram is a good way to compare two sets of data. Find out how to draw and interpret a back-to-back stem and leaf diagram.

 THINGS TO DO AND THINK ABOUT

Use the stem and leaf diagram above to answer the following question.

1. Students were each given a position according to their marks. Iona was 1st with 89 marks. What position was William who scored 54 marks?

MORE TABLES AND GRAPHS

COMPOUND BAR GRAPHS

A **compound bar graph** is an extension of an ordinary bar graph which compares two or more quantities at the same time. The graph shown below was produced by the Scottish Government in 2012 in their publications and compares the median gross weekly earnings of *all* full-time workers, *female* full-time workers and *male* full-time workers for the decade 2001–2010.

EXAMPLE

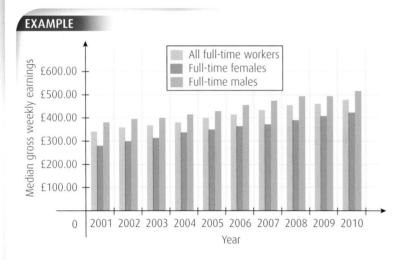

(a) What was the trend of weekly earnings over the period 2001–2010?

(b) Compare the weekly earnings of full-time female workers with those of full-time male workers over the period 2001–2010.

(c) It was claimed that, in the period 2003–2010, the gross weekly earnings of full-time male workers increased by more than 25%.
Comment on this claim.

SOLUTION

(a) The trend of the graph shows that weekly earnings were increasing during the period 2001–2010.

(b) Throughout this period, full-time female workers earned less than full-time male workers.

(c) In 2003, the median gross weekly earnings of full-time male workers was £400. As 25% of £400 = £400 ÷ 4 = £100, we should check the graph to see if the median is £400 + £100 = £500 in 2010. As the median gross weekly earnings of full-time male workers was just over £500 in 2010, we can say the claim is correct.

DON'T FORGET

The **trend** of a graph indicates the general direction that the graph is going in, for example increasing (upwards), decreasing (downwards) or stable (not changing). If you are asked to compare or comment, remember to answer in words.

MILEAGE CHARTS

Distances between towns and cities are sometimes displayed in a mileage chart. The mileage chart below shows the distances, in miles, between Scotland's seven cities.

Aberdeen						
67	Dundee					
125	63	Edinburgh				
145	83	46	Glasgow			
105	138	158	171	Inverness		
87	22	44	60	113	Perth	
123	58	38	30	145	34	Stirling

To read the chart, find where a column and a row intersect, for example the distance from Dundee to Inverness is 138 miles.

(a) What is the distance between Dundee and Perth?
(b) Which two cities are 158 miles apart?
(c) Denis is a delivery driver. He has to drive from Edinburgh to Stirling, then on to Perth and finally return to Edinburgh. How far will he drive?

(a) 22 miles
(b) Edinburgh and Inverness
(c) (38 + 34 + 44) miles = 116 miles

MAKING DECISIONS BASED ON A GRAPH

The graph below compares the annual interest rates offered by a building society on its Loyalty and Champion accounts.

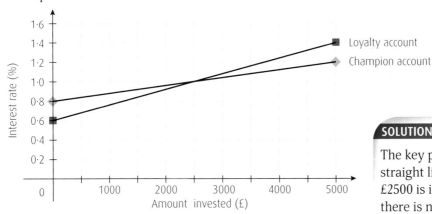

Lucy intends to invest money in one of these accounts. What advice would you give her?

The key point on the graph is where the two straight lines intersect. At that point, where £2500 is invested and the interest rate is 1%, there is no difference between the accounts. Therefore sensible advice to Lucy would be to invest in a Champion Account if she has less than £2500, but to invest in a Loyalty Account if she has more than £2500.

 COURSE IDEA

If you visit a bank or building society, you will be able pick up leaflets with information about different types of accounts such as current accounts (which come with a cheque book), individual savings accounts or ISAs (which are tax free), instant access accounts, fixed rate bonds and so on. Try to understand the different interest rates and conditions which apply to these accounts.

 THINGS TO DO AND THINK ABOUT

Use the earlier tables and graphs in this section to answer the following questions.

1. What were the median gross weekly earnings of a full-time female worker in 2002?

2. How much interest would be earned on a sum of £5000 invested for 1 year in a Loyalty Account?

3. Jack is going to drive from his home in Inverness to Perth to see Inverness Caledonian Thistle play St. Johnstone in the league. The kick-off is at 3.00pm. He leaves Inverness at 11.45am and drives at an average speed of 40 miles per hour. Will he arrive in time for the kick-off? Justify your answer by calculation.

EXTRA SECTIONS

PRACTICE ADDED VALUE UNIT TEST

PAPER ONE

Do not use a calculator. Allow 20 minutes to do Paper One.

1. In a shop, in order to buy a fridge on hire purchase, you must pay a deposit of 15% of the marked price. How much would the deposit be on a fridge with a marked price of £480?

2. A golfer records the following scores in six rounds of golf.
   72   75   78   71   70   73
   Find his mean score. Give your answer correct to 2 decimal places.

3. A factory employs 180 people. In a survey it is found that $\frac{2}{5}$ of the employees travel to work by train. How many of the employees travel to work by train?

4. Joseph has a plank of wood which is 9·5 metres long. He cuts it into three pieces. One piece is 2·63 metres long. Another piece is 4·81 metres long. What is the length of the third piece?

5. Ahmad orders a magazine each month. It costs £4·79. How much will it cost Ahmad to buy six of these magazines?

DON'T FORGET

Before you start PAPER TWO, remember to check that your calculator is in the correct mode for trigonometry. If tan 45° = 1, then you are in degree mode and that is correct.

PAPER TWO

You may use a calculator. Allow 40 minutes to do Paper Two.

1. Solve algebraically the equation
   $5y - 1 = 2y + 20$

2. Kylie buys a fish tank which is in the shape of a cuboid.
   The fish tank is 75 centimetres long 28 centimetres broad and 32 centimetres high.

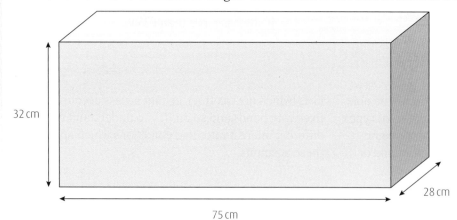

32 cm

28 cm

75 cm

   Calculate the volume of the fish tank in litres.

3. A garden fence is constructed by joining sections. The sections are made of metal girders.

1 section	2 sections	3 sections

 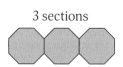

   8 girders   15 girders

(a) Complete the table below.

Number of sections (s)	1	2	3	4		10
Number of girders (g)	8	15				

(b) Write down a formula for calculating the number of girders (g) when you know the number of sections (s).

(c) 106 girders are used to construct a garden fence. How many sections are in the fence?

4. Madeleine sets off on a journey on the motorway at 0950. She drives at an average speed of 104 kilometres per hour. She stops for a break at a service station at 1305. How far has Madeleine driven?

5. A side view of a lean-to greenhouse is shown in the diagram.
Calculate the length of the sloping part of the roof.

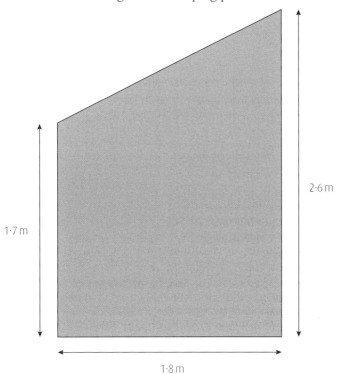

2·6 m

1·7 m

1·8 m

6. A ramp is designed to allow goods to be loaded into the back of a truck.
The length of the ramp is 250 centimetres.
The horizontal distance of the ramp is 220 centimetres.

Calculate the size of the angle marked $x°$.

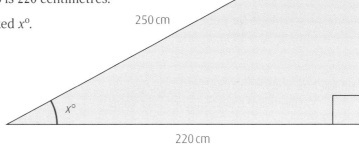

250 cm

$x°$

220 cm

7. (a) On a coordinate grid, Plot the points P (5, 2), Q (3, –5) and R (–4, –3).

(b) Plot a fourth point S to form a square PQRS.

(c) Write down the coordinates of S.

WORKED SOLUTIONS TO ADDED VALUE UNIT TEST

PAPER ONE

1. 10% of £480 = £48, so 5% of £480 = £48 ÷ 2 = £24
   Hence 15% of £480 = £48 + £24 = £72

2. The mean = $\frac{\text{Total of all values}}{\text{Number of values}} = \frac{72 + 75 + 78 + 71 + 70 + 73}{6} = \frac{439}{6} = 439\cdot000 \div 6 = 73\cdot166$
   Hence mean score = 73·17 (correct to 2 decimal places).

3. $\frac{2}{5}$ of 180 = 180 ÷ 5 × 2 = 36 × 2 = 72, hence 72 employees travel by train.

4. Length of third piece = 9·5 – (2·63 + 4·81) = 9·50 – 7·44 = 2·06
   Hence the third piece is 2·06 m long.

5. Cost of magazines = 6 × £4·79 = £28·74

PAPER TWO

1.

	5y	–	1	=	2y	+	20	
			+1				+1	(add 1 to each side of the equation)
⇒			5y	=	2y	+	21	
			–2y		–2y			(subtract 2y from each side of the equation)
⇒			3y	=	21			
			÷3		÷3			(divide each side of the equation by 3)
⇒			y	=	7			

2. $V = lbh = 75 \times 28 \times 32 = 67\,200$, hence volume = 67 200 cm³.
   Volume in litres = (67 200 ÷ 1000) litres = 67·2 l.

3. (a) By counting 3 → **22**, then by adding on 7 each time 4 → **29**, 5 → 36, 6 → 43,
   7 → 50, 8 → 57, 9 → 64, 10 → **71**.

Number of sections (s)	1	2	3	4		10
Number of girders (g)	8	15	**22**	**29**		**71**

   (b) As we are adding on 7 each time, first step in formula is × 7, then we must add
   1 for the second step so formula is $g = 7s + 1$

   (c) Substitute s = 106 in the formula, leading to 7s + 1 = 106, subtract 1 from
   each side leading to 7s = 105 then divide each side by 7. The solution is then
   s = 105 ÷ 7 = 15

4. 0950 → 1050 → 1150 → 1250 = 3 hours
   1250 → 1300 = 10 minutes *and* 1300 → 1305 = 5 minutes
   Hence journey time is 3 hours + (10 + 5) minutes = 3 hours 15 minutes.
   15 minutes = 15 ÷ 60 = 0·25 hours, so 3 hours 15 minutes = 3·25 hours.
   $D = ST = 104 \times 3\cdot25 = 338$
   Hence Madeleine has driven 338 km.

5. Create a right-angled triangle using a dotted line.
    Check that the vertical side in the right-angled triangle = (2·6 – 1·7) m = 0·9 m.
    Let the sloping part of the roof = x metres.

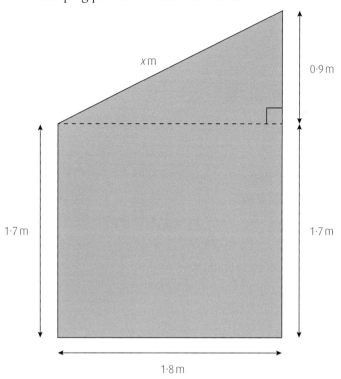

Using the Theorem of Pythagoras,
$x^2 = 1\cdot8^2 + 0\cdot9^2 = 3\cdot24 + 0\cdot81 = 4\cdot05$

Hence $x = \sqrt{4\cdot05} = 2\cdot01246118$, so the sloping part of the roof = 2·0 m (correct to 1 decimal place).

6. The given sides are the *adjacent* and the *hypotenuse*. Check the formula list for
    adjacent and hypotenuse you will see that we must use cosine.

    $\cos x^o = \dfrac{\text{adjacent}}{\text{hypotenuse}} = \dfrac{220}{250}$

    Now press SHIFT cos (220 ÷ 250) =
    This will lead to $x = \cos^{-1}(220 \div 250) = 28\cdot35763658$
    Hence the marked angle $x^o = 28°$ (to the nearest degree).

7. (a)

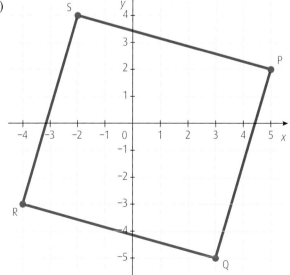

    (b) S plotted correctly.

    (c) (–2, 4)

DON'T FORGET

You may think that the times allowed for the two test papers, that is 20 minutes for Paper One and 40 minutes for Paper Two, were very generous. However, there is a good reason for this. The time is given in order that you can check and re-check all your calculations, check that you have copied all the figures from the test papers correctly, check that you have not missed anything out and check any diagrams required. It is essential that you carry out all these checks.

MORE PRACTICE

In this section, you will get the opportunity to practise extra examples covering *some* of topics from each of the three units of the course. The examples chosen would seem those likely to be included in unit assessments. You may use a calculator for all three units.

QUESTIONS FROM EXPRESSIONS AND FORMULAE

1. (a) Expand the brackets and simplify: $5p + 3(p - 4)$

    (b) Factorise $14a - 35$

2. If $x = 7$ and $y = 3$, find the value of $6x - 5y$.

3. The perimeter of a rectangle can be found using the formula $P = 2(l + b)$.
    Calculate b when $P = 36$ and $l = 11$.

4. Calculate the gradient of the straight line joining the points (4, 1) and (6, 7).

5. The centre circle on a basketball court has a radius of 1·83 metres. Find the circumference of the centre circle. Give your answer correct to one decimal place.

6. The area of the base of a cylindrical container is 2500 square centimetres. If the height of the container is 80 centimetres, calculate the volume of the container. State the units with your answer.

7. The heights, in centimetres, of a group of students are listed below.

    156   162   154   155   149   159   161   160

    Calculate (a) the mean height   (b) the median height   (c) the range.

8. Complete this shape so that it has rotational symmetry of order 4 about O.

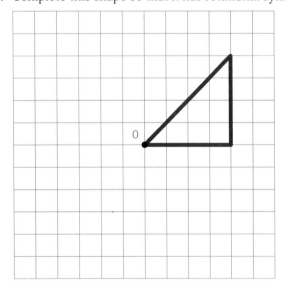

QUESTIONS FROM RELATIONSHIPS

1. (a) Complete the table below for $y = 3x + 1$

x	1	2	3
y			

    (b) Draw the line $y = 3x + 1$

2. Solve the following equation: $5x + 4 = -11$.

3. Change the subject of the formula $p = qx - r$ to x.

4. Calculate the distance between the points $(1, -7)$ and $(6, 5)$.

5. A motorcyclist travels 84 miles in 1 hour 30 minutes.
   Calculate his speed in miles per hour.

6.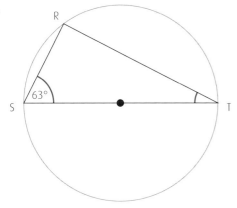

ST is a diameter of the circle.
Point R lies on the circumference of the circle.
Angle RST = 63°.
Calculate the size of angle RTS.

QUESTIONS FROM NUMERACY

1. The bill in a restaurant comes to £52. If a service charge of 15% is added to the bill, find the total cost of the bill.

2. Dean returns from a skiing holiday in Switzerland with 45 Swiss francs. Convert this amount to pounds sterling if the exchange rate is £1 = 1·47 Swiss francs. Give your answer correct to the nearest penny.

3. £250 has to be divided between Mark and Gareth in the ratio 3:7. How much will Mark receive?

4. Giovanni earns £78 from a part-time job one week. He saves 2/5 of his earnings. How much does he save?

5. In a survey, 150 people were asked to name their favourite colour. 57 of those surveyed said blue was their favourite colour. What percentage of those surveyed was this?

6. The cost of eight copies of a newspaper is £10·40. How much would three copies of this newspaper cost?

7. Bradley has bought tickets for two different raffles.

   In Raffle A, 500 raffle tickets are sold and Bradley buys 21 of them.
   In Raffle B, 600 raffle tickets are sold and Bradley buys 24 of them.

   In which raffle does Bradley have a better chance of winning a prize.
   Justify your answer with your calculations.

8. Marie sets out in her car at 09.45. She drives 180 kilometres at an average speed of 40 kilometres per hour to her destination. When will she arrive at her destination?

SOLUTIONS TO THINGS TO DO AND THINK ABOUT 1

The following two sections contain solutions to the questions in the 'THINGS TO DO AND THINK ABOUT' part of each topic as well as solutions to the questions in the 'MORE PRACTICE' section.

SOLUTIONS TO EXPRESSIONS AND FORMULAE

Page 7: 1(a) $7m$ (b) $4p$ (c) $8x$ (d) $15q$ 2(a) $4m + 4$ (b) $3x - 12$ (c) $16a + 40$
(d) $10 - 30z$ 3. 12 4(a) $5(b + 6)$ (b) $4(3a - 4)$ (c) $7(y + 8)$ (d) $6(5 - 6c)$

Page 9: 1(a) $11y$ (b) $8p + 2q$ 2(a) $14x + 20$ (b) $7x + 9$
3(a) 22 (b) 10 (c) 65 (d) 3

Page 11: 32·4

Page 13: (a)

Number of sections (s)	1	2	3	4		11
Number of lengths of wood (w)	7	13	19	25		67

(b) $w = 6s + 1$ (c) 23

Page 15: 1. No, 0·0666... > 0·05 2. $\frac{1}{2}$

Page 17: 1. 198 cm 2. 69 cm 3. 36 cm 4. 126 cm

Page 19: 1(a) 57·5 m (b) 263 m² 2. 5·9 m²

Page 21: 1170 cm²

Page 23: 1. 7 faces, 7 vertices, 12 edges 2. Hexagonal based pyramid 3. 168 cm

Page 25: 1. 864 cm² 2. 236 cm² 3. 603 cm²

Page 27: 1. 600 m³ 2. 8 cm³ 3. 165 cm³

Page 29: 1.

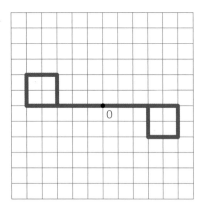

2.

Page 31: 1.

Marks	Tally	Frequency
30 – 39			1
40 – 49	||||	4
50 – 59	|||| ||||	10
60 – 69	|||| |||	8
70 – 79	||||	5
80 – 89	||	2

2 (a) discrete   (b) continuous   (c) continuous   (d) discrete

Page 33: mean = 56·2, median = 58·5, mode = 63, range = 31

Page 35: 162°

Page 37: $\frac{2}{11}$

SOLUTIONS TO RELATIONSHIPS

Page 39: (a)

x	1	2	3	4
y	1	3	5	7

(b)
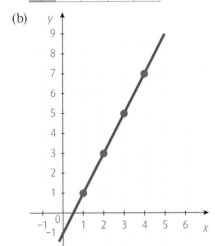

Page 41: 1(a) $x = 2$ (b) $x = 8$ (c) $x = 6$ (d) $x = -2$ 2. $y \leqslant 5$

Page 43: 1. $G = N + D$ 2. $I = \frac{V}{R}$ 3. $Y = 12M$ 4. $c = \frac{s-1}{4}$ 5. $y = \frac{A+Q}{P}$

Page 45: 53 cm
Note: The missing Pythagorean triples are (5, 12, 13), (8, 15, 17), (12, 16, 20), and (7, 24, 25).

Page 47:
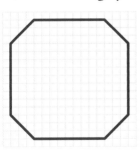

Page 49: 84°

Page 51: 102°

Page 53: 1(a) 12·6 cm   (b) 16·8 cm   2. 54 m

Page 55: 39·5°

Page 57: 1. 36 (±1)   2. 67 (±1)

SOLUTIONS TO THINGS TO DO AND THINK ABOUT 2

SOLUTIONS TO NUMERACY

Page 59: No, 360 ml > 350 ml

Page 61: 1. 6·01   2. 6·0   3. 17   4. 6790   5. 570 cm²

Page 63: 1(a) 5·27   (b) 68·22   (c) 562·7   2. £5·26

Page 65: 1. $\frac{3}{5}$ 2. 0·625   3. 36   4. 24

Page 67: 1. $\frac{3}{5}$ 2. 19 kg   3. 225   4. 64%

5. English because 0·71875 > 0·7 and 0·68

Page 69: 1. £16 362   2. £1200   3. £12   4(a) $664   (b) £30·12   5. £14 150

Page 71: 1. 0·135 l   2. 15 l   3. 5·6

Page 73: 1. 78 euros   2. 470 g

Page 75: 1(a) 6   (b) 4   (c) 0   (d) –9   (e) 6   2(a) –21   (b) 0   (c) 1   (d) –4   3. 17

Page 77: 1. 4 h 38 m   2. 0221

Page 79: 1. 245 miles   2. 42 km

Page 81: £174·65

Page 83: 20th

Page 85: 1. £300   2. £70   3. Yes, he could arrive in time for kick-off; he could drive 130 miles and it is only 113 miles from Inverness to Perth (or equivalent)

SOLUTIONS TO MORE PRACTICE

EXPESSIONS AND FORMULAE

1. (a) $8p – 12$ (b) $7(2a – 5)$

2. 27

3. $b = 7$

4. 3

5. 11·5 m

6. 200 000 cm³ or 200 l

7. (a) 157 cm   (b) 157·5 cm   (c) 13 cm

8.

RELATIONSHIPS

1. (a)

x	1	2	3
y	4	7	10

(b)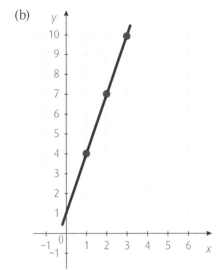

2. $x = -3$

3. $x = \frac{p + r}{q}$

4. 13 units

5. 56 mph

6. 27°

NUMERACY

1. £59·80

2. £30·61

3. £75

4. £31·20

5. 38%

6. £3·90

7. He will have a better chance of winning a prize in raffle A because $\frac{21}{500} = 0.042$ as a decimal fraction, whereas $\frac{24}{600} = 0.04$ as a decimal fraction and $0.042 > 0.04$.

8. 1415 or 2.15pm

ADVICE:

As you check through the solutions to the questions in the 'THINGS TO DO AND THINK ABOUT' and 'MORE PRACTICE' sections, it is important that you take action over any questions which you got wrong. Try to find out the cause of any errors with a view to eliminating similar errors in future.

A – Z GLOSSARY OF MATHEMATICAL TERMS

acute:
an acute angle is less than 90º.

alternate angles:
angles which occur when a straight line intersects parallel lines. They are equal and form a Z-shape.

circumference:
the distance around the outside of a circle, found using the formula $C = \pi d$.

class intervals:
when arranging widely spread data into a frequency table, class intervals such as 10 – 19, 20 – 29, 30 – 39 and so on make the task simpler.

commission:
a way of rewarding salespersons by giving them a percentage of their sales as part of their salary.

composite:
a composite shape is a shape made up of two or more shapes, for example a rectangle and a triangle.

compound bar graph:
an extension of an ordinary bar graph which compares two or more quantities at the same time.

continuous:
continuous data is measured data which can take any value in a given range, for example the heights of a group of students.

correlation:
correlation describes the connection between two sets of data, for example height and weight.

corresponding angles:
angles which occur when a straight line intersects parallel lines. They are equal and form an F-shape.

denominator:
the bottom number in a fraction, for example 5 in $\frac{3}{5}$.

direct proportion:
two quantities are said to be in direct proportion if, as one quantity increases or decreases, the other increases or decreases at the same rate.

discount:
a discount reduces the price of an item in a sale.

discrete:
discrete data is data which can only take certain values in a given range, for example the number of goals scored in a football match.

equation:
an equation is a sentence with the verb 'is equal to' in it.

expression:
an expression is a term such as $2x$ or a collection of terms such as $2x + 3y$.

gradient:
can be found using the formula Gradient = $\frac{\text{Vertical height}}{\text{Horizontal distance}}$ and tells you the slope of a line.

highest common factor (HCF):
the HCF of two or more numbers is the largest factor common to these numbers, for example the HCF of 20 and 30 is 10.

hire purchase:
a way of buying expensive items by paying a small part of the cost, called the deposit, followed by monthly instalments.

hypotenuse:
the longest side in a right-angled triangle. It is opposite the right angle.

inequation:
an inequation is a sentence containing 'is greater than' or 'is less than' in it.

integer:
the set of integers {...–3, –2, –1, 0, 1, 2, 3...} is the set of positive and negative whole numbers and zero.

interest:
interest is paid by banks to its customers as a percentage of the amount they have deposited in an account.

isosceles triangle:
a triangle with two equal sides and angles.

like terms:
terms such as $5a$ and $2a$ are like terms and $5a + 2a$ can be simplified leading to $7a$.

line symmetry:
where one half of a shape is the reflection of the other half in an axis of symmetry.

mean:
an average of a set of data defined by $\frac{\text{Total of all values}}{\text{Number of values}}$.

median:
the middle value in a set of ordered values.

mode:
the most frequent value in a data set.

net:
the net of a solid shape is a flat two-dimensional shape which can be cut out and folded up to make the solid shape.

numerator:
the top number in a fraction, for example 3 in $\frac{3}{5}$.

obtuse:
an obtuse angle is greater than 90º and less than 180º.

perpendicular:
two lines are perpendicular if they are at right angles to one another.

prime number:
a number that is only divisible by 1 and itself.

prism:
a prism is a solid shape which has opposite ends that are parallel and congruent, for example a cuboid.

probability:
a measure of how likely an event is to happen.

quadrilateral:
any four-sided figure such as a square, kite, rhombus, etc.

range:
for any numerical data set, the range = highest value – lowest value.

reflex:
a reflex angle is greater than 180º and less than 360º.

rotational symmetry:
a shape has rotational symmetry when it fits its outline as it turns (or rotates).

scale factor:
the size of an enlargement (or reduction) is described by its scale factor. For example, a scale factor of 2 means that all the lengths in the new shape are twice the lengths of those in the original shape.

scattergraph:
A scattergraph is a statistical diagram which can be used to compare two sets of data by plotting a set of points on a coordinate grid.

similar:
two shapes are said to be similar when one shape is an enlargement or reduction of the other.

stem and leaf diagram:
an ordered statistical diagram in which each member of the data set is split into a 'stem' and a 'leaf'.

tangent:
a tangent to a circle is a straight line which touches the circle at one point only. This point is called the point of contact.

Theorem of Pythagoras:
states that in a right-angled triangle, the square on the hypotenuse is equal to the sum of the squares on the other two sides, usually given by the formula $a^2 + b^2 = c^2$.

trend:
the trend of a graph indicates the general direction that the graph is going, for example increasing or decreasing.

value added tax:
value added tax, or VAT for short, is added to the cost of certain goods.

vertically opposite angles:
vertically opposite angles are equal. They are formed when two straight lines intersect.